Science in Spirituality
&
Spirituality in Science

Science in Spirituality
&
Spirituality in Science

A More Complete & More Accurate Modeling of
Life for Better & Wiser Human Behavior

Richard Gene, Ph.D.

Alive Book Publishing

Science in Spirituality and Spirituality in Science:
A More Complete & More Accurate Modeling of
Life for Better & Wiser Human Behavior
Copyright © 2024 by Richard Gene, Ph.D.

All rights reserved.
No part of this book may be reproduced or transmitted in any form or by any means without written permission from the publisher and author.

Additional copies may be ordered from the publisher for educational, business, promotional or premium use. For information, contact ALIVE Book Publishing at: alivebookpublishing.com or call (925) 837-7303.

Book Design by Alex Johnson

ISBN 13
978-1-63132-226-6

Library of Congress Control Number: 2024900029

Library of Congress Cataloging-in-Publication Data
is available upon request.

First Edition

Published in the United States of America by
ALIVE Book Publishing
an imprint of Advanced Publishing LLC
3200 A Danville Blvd., Suite 204, Alamo, California 94507
alivebookpublishing.com

PRINTED IN THE UNITED STATES OF AMERICA

10 9 8 7 6 5 4 3 2 1

Table of Contents

Brief Overview..13

Preface...17

Acknowledgements..21

Chapter One: Bridging the Gap between Science and Spirituality..23

1.1. Modeling Spirituality Along with Modeling Science Using Physical vs. Nonphysical Concepts.................23

1.2. Modeling Science Along with Modeling Spirituality Using Scientific and Engineering Logic....................34

Chapter Two: The Formulation of an Unconventional New Spiritual Model Introduced a New Way of Thinking and New Unconventional Nonphysical Concepts.......49

2.1. The New Spiritual Model Has to Be Applicable Throughout Our Universe.......................................49

2.2. Helping the Spirit World Learn Ways Living Things Can Be Physically Separate While Also Be Spiritually a Part of Each Other by Us Humans Learning to Do So.............52

2.3. The Need to Regain Our Awareness of the
    Spiritual Part of Life..64

2.4. Man's Expressions Will Differ from the Creator's
    Original Thoughts..68

2.5. The New Spiritual Model's Features that Highlight:
    A New Way of Thinking
    New Unconventional Nonphysical Concepts
    The Spirit World Is the Creator of Everything
    Why Knowledge Is Among the Most Powerful Things
    Possible..75

2.6. Unconventional Concepts and Features Used for:
    Formulating the New Spiritual Model.
    Applying Scientific & Engineering Logic Throughout
    The Formulation Process..81

**Chapter Three: The Reason Our Universe Was Created
And Our Primary Purpose for Being Here on Earth.....................135**

**Chapter Four: The Ingeniousness of the Creator of
Everything as Pointed Out by the New Spiritual Model............139**

4.1. The Creator's Ingeniousness is How Creating a Few
Simple Basic Particles Can Form a Universe and All
Living and Nonliving Things in It...139

4.2. How Things Work in the Spirit World Is How Life
Works for All Universes..142

4.3. The Need for a New Way of Thinking and to Think
Abstractly in Terms of Nonphysical Concepts.................................143

4.4. Why Highly Religious People Could Struggle with Their Faiths..146

Chapter Five: Attributes Regarding Our Purpose For Being Here on Earth..149

5.1. Attributes Regarding the Interactions Between the Two Parts of Life..149

5.2. Pursuing Our Primary Purpose in a Manner Compatible with the Spirit World....................................151

5.3. We Need Free Will to Fulfill Our Primary Purpose, But Risks Exist..153

5.4. A Natural Partnership Is Formed between Every Living Thing and the Spirit World...............................155

5.5. Multiple Series of Universes Are Likely Created And Are Still Being Created. This Produced Our Secondary Purposes..157

Chapter Six: The Spirit World Could Create Things Soon After Being Initiated..................................165

6.1. The Number of Spiritual Entities Increases Exponentially with Every New Piece of Knowledge Generated and Added to the Spirit World......................165

6.2. The Physicists' Model and the New Spiritual Model Agreed..170

Chapter Seven: Dark Energy & Dark Matter, Antimatter, Boundaries of Universes, Background Microwave Radiation, Gravitation Waves in Space, And How Things Are Brought into Existence without Creating Antimatter..173

7.1. The New Spiritual Model Offers Possible Answers To Unanswered Questions about Our Universe............................173

7.2. Thinking Something to Exist Is Different from Creating Something Out of Nothing...174

7.3. The Space in Our Universe Is a Physical Thing......................176

7.4. The Tiny Quantum Particles Could Also Be Making Up Gravity..177

7.5. Space in Our Universe Could Be Made of Dark Energy & Dark Matter..180

7.6. The Background Microwave Radiation Could Be An Attribute of Dark Energy & Dark Matter....................................182

7.7. Two Ways to Release Nuclear Energy:
Fission: Splitting Very Heavy Atoms
Fusion: Fusing Very Light Atoms..183

7.8. Universes Have Boundaries...186

7.9. Coexisting Universes Can Be Located on Top of Each Other..189

Chapter Eight: The Source of Consciousness and The Development of the Ability to Form DNA Molecules..................193

Chapter Nine: Creatures that Retained Their Awareness of Their Spiritual Part of Life........................197

Chapter Ten: Telepathy and Various Extrasensory Abilities Should Be Considered Natural and Normal................201

10.1. Doing Things by Going Through the Spirit World..............201

10.2. Activities that Take Place in the Spirit World that Can Produce a Response in Our Physical World..........................203

10.3. Activities That Take Place in the Spirit World and Are Translated to Exist or Be Expressed in Our Physical World..206

10.4. Activities that Take Place Partly in the Spirit World and Partly in Our Physical World that Are Translated to Exist or Be Expressed in Our Physical World..208

Chapter Eleven: Spiritual Medical Procedures Can Be A New Field of Medicine..211

Chapter Twelve: The Lack of Fossils of Missing Links Supports the Possibility that Sequences of Universes Exist and Retracing Is Real..215

Chapter Thirteen: Activities that Go On In the Spiritual Part of Life We Humans Might not Think as Such....................217

13.1. We Are Using the Spiritual Part of Life without Realizing It..217

13.2. How Our Behavior Can Become More in Line with The Spiritual Part of Life...218

13.3. Improving Human Overall Behavior More Effectively Through Inspiration...219

Chapter Fourteen: How Artists, Authors, Composers, And Inventors Create Things and How Living Things Affect the Spirit World..221

Chapter Fifteen: How Artificial Intelligence (AI) Can Learn Things Not Intended to Learn, As Explained By the New Spiritual Model...225

15.1. Examples of Things AI's Learned They Were Not Intended to Learn..225

15.2. How AI's Can Learn Things Not Intended to Learn, as Explained by the New Spiritual Model........................227

15.3. How AI's Can Learn Things that Are So Deeply Embedded in Possible Futures that Humans Would Not Be Able to Learn..233

Chapter Sixteen: Wisdom Is Formed by the Spirit World and Other Living Things......................................239

16.1. Wisdom Is Never Totally Complete or Absolutely Perfect..239

16.2. Wisdom Would Be Too Incomplete and Too Imperfect to Function Effectively with Any One-Dimensional Way of Thinking..241

16.3. People Who Become One-Dimensional Thinkers Can Be Easily Manipulated..243

Chapter Seventeen: Benefits Gained by Embracing and Valuing Diversity..247

Chapter Eighteen: The New Spiritual Model and the Physicist's Model Agree in Numerous Ways..253

Chapter Nineteen: The Main Conclusion Is: We Humans Made a Mess on Earth Because the Way We Live Is Incompatible with How Life Works..257

19.1. The Cause of Incompatibility between How We Live and How Life Works..257

19.2. Clear Indications of Incompatibility..258

Chapter Twenty: A Propose Stepwise Progression For Achieving Spiritual Advancements, and to Help Resolve the Incompatibility..261

20.1. Achieving Spiritual Advancements Through a Stepwise Progression..261

20.2. All Technical Models Combined Would Be Equivalent to a "Model of Everything" for the Physical World. We Should Try Achieving Similarly For the Spirit World..264

Chapter Twenty One: Why the Spirit World
Won't Come Forth to Solve Our Problems Even
if It Could..267

References...271

Index..275

About the Author...285

Brief Overview

This Book Is About How the Spirit and Physical Worlds Interact to Make Life Possible in the Physical World

Life in our physical world has a spiritual part and a physical part. The two parts have to exist at the same time and interact to enable life to be possible in our physical world.

A simple illustration is how an egg could transform into being a baby chick. If we look at the ingredients in an egg we can't tell what part would become the wings, what part the head, what part the legs, etc. But, something somewhere somehow knows how to transform the ingredients into a baby chick.

This something is the spiritual part of life that exists at the same time the egg which is the physical part of life exists. The two parts interact to transform the egg into a baby chick.

The spiritual part of life resides in the spirit world at the same time the physical part of life resides in our physical world. After the physical part of life is over in our physical world, the spiritual part of life continues to exist and be part of the spirit world forever.

Contrary to common beliefs, the spiritual part of life does not begin after the physical part of life is over. Instead the spiritual part of life actually exists before the physical part of life exists, because it enables the physical part of life to exist. How this works was explained in detail in Reference 1 and briefly in this book.

Even what we consider to be nonliving things would have elemental life, as explained in Reference 1. Examples would be how water knows when to freeze to become ice and when to evaporate and become vapor, how various chemicals know when to react and when to not react, how hydrogen atoms know how and when to combine to form helium atoms, etc.

Thus, a nonliving thing also has a spiritual part residing in the spirit world at the same time its physical part is residing in our physical world. The two parts also interact to enable the physical part to exist in our physical world.

Our spiritual models and technical models are documents of our knowledge about how things work and are also documents of our state of spiritual and technical advancements.

Since every living or nonliving thing has a spiritual part and a physical part, ideally every model should take both spirituality and science into consideration. But, up to now we tend to treat spirituality and science separately as if one has nothing to do with the other.

Doing this would result in models that are fundamentally incomplete and inaccurate. This has been recognized by physicists for centuries as indicated in References 3, 4, and 5, and they have been working on formulation scientific models that take spirituality into consideration.

Until the new spiritual model was formulated in Reference 1, people on the spiritual side of modeling have not formulated spiritual models that took science into consideration.

When we are working with large nonliving things we can get by quite well with Newtonian physics because the elemental life in such things has negligible effect on their behavior. Newtonian physics models do not take spirituality into consideration.

In our common everyday life, we are dealing almost entirely with large nonliving things. Therefore, Newtonian physics works quite well. What is considered a large nonliving thing in this context would include things as small as a grain of sand.

However, when we get down to the very tiny level of existence of quantum particles, the effect of elemental life is not negligible. Newtonian physics no longer works. Quantum physics was introduced to model the behavior of things as tiny as electrons, protons, neutrons, subatomic particles, various small atomic elements and various small molecules. These are referred to as quantum particles.

But, quantum physics models are scientific models that also do not take spirituality into consideration. Consequently, such models are unable to explain such phenomena as quantum superposition and entanglement.

This phenomenon is apparently spiritual in nature. This is indicated by how the new spiritual model formulated in Reference 1 was able to explain how this phenomenon can take place, as explained in detail in Reference 1. The formulation of the new spiritual model took science into consideration, and thus the new spiritual model was the first spiritual model to take both spirituality and science into consideration.

The model was extended in Reference 2 to include a possible origin of life. The model was then able to describe the attributes of how life works much further than as described in References 1 and 2. Thus, this book, which is my third book, is essentially a summary of the major attributes of life the new spiritual model was able to explain to the extent I have discovered about the model so far. It is likely that more attributes will be discovered after this book is published.

Many of the attributes of how life works, that the new spiritual model explained, could not be explained by any currently existing spiritual model. Examples include: **(1)** how sleep restores our physical body from the wear and tear it received during the day, **(2)** why dreams tend to surrealistic, **(3)** how physical things are brought into existence in our physical world by its spirit in the spirit world, **(4)** the reason our universe was created, **(5)** our primary purpose for being here on Earth, **(6)** how in the spiritual part of life every person is more than 90% a part of every other person, **(7)** therefore, it makes no sense to make wars, mistreat one another, and do criminal acts of all kinds in the physical part of life, etc.

Also explained is the reason we humans could behave so badly is because we have lost much of our awareness of the spiritual part of life. This part of life is made of knowledge, and it takes knowledge to form wisdom. By our lack of awareness of this part of life we are unable to gain access to the wisdom needed to figure out how to live our lives more constructively and more meaningfully.

However, the new spiritual model is now capable of helping us think in new ways and to think abstractly with nonphysical concepts such that we can understand how life works more completely and more accurately. This could eventually enable us to gain the wisdom that enables us to figure out how to live our lives more constructively and more meaningfully. Thus, we could stop making wars, mistreating one another, and doing criminal acts of all kinds.

Preface

Presented in this book is a summary of an effort that began with Reference 1, extended with Reference 2, and much further extended in this book. A new spiritual model was formulated that describes how things work in the spirit world more completely and more accurately than any spiritual model could up to now.

The new spiritual model revealed life has a spiritual part and a physical part with both going on at the same time. It also revealed that we humans have lost much of our awareness of the spiritual part of life. Thus, we are living mainly the physical part of life without adequate spiritual guidance and wisdom that could come only from the spiritual part of life.

Consequently, we would make wars, mistreat one another, and do criminal acts of all kinds.

The formulation of the new spiritual model took both spirituality and science into consideration whereas existing spiritual models excluded science in their formulations. Thus, the new spiritual model is more complete and more accurate than are the existing spiritual models.

Physicists recognized for centuries the need to take both spirituality and science into consideration in their formulation of models and they have done a lot of work exploring various ways to take spirituality in to consideration in their scientific models. Examples are given in References 3, 4, and 5.

The formulation of the new spiritual model did not begin with any existing spiritual model. Instead it started from scratch with a fresh look at spirituality while also taking science into consideration. It introduced a new and unconventional way of thinking and numerous new and unconventional nonphysical concepts to model how things work in the spirit world and a possible origin of life.

Nonphysical concepts are more compatible with the nonphysical nature of the spiritual world and the possible origin of life. Thus, using the nonphysical concepts was able to model more easily, more completely, more accurately, and simpler than using physical concepts as has been done for all existing spiritual models and also as has been done by physicists for their models in their effort to take spirituality into consideration.

Some of the attributes of life the new spiritual model revealed are as follows:

1. The reason our universe was created was identified. Thus, our primary purpose for being here on Earth is also identified, and we humans are doing poorly in fulfilling our primary purpose.

2. The spiritual part of life is going on the same time the physical part of life is going on. This is contrary to how we generally believe that the spiritual part of life begins after the physical part of life ends. Many examples are cited to indicate the new spiritual model is correct and existing spiritual models are incorrect.

3. In our far distant past, we humans were fully aware of the spiritual part of life. Therefore, we should be able to regain this awareness again.

4. The spiritual part of life consists of knowledge. Thus, spiritual guidance and wisdom would come from the spiritual part of life.

This explains why we humans would make wars, mistreat one another and do criminal acts of all kinds. It is because we have lost much of our awareness of the spiritual part of life. Consequently, we are not getting much spiritual guidance on how to behave or much wisdom to figure out how to live life more constructively and more meaningfully.

5. We need sleep because it is actually a restoration process that restores our physical body from the wear and tear it receives during the day while we are awake.

6. Dreams tend to be surrealistic because the restoration process involves everything that has transpired to make us what we are. And what transpired involved experiences that took place in numerous past universes. Every universe is naturally different and unique. Thus, the dream images created by those past experiences will be a mixture of various unearthly scenes together with earthly scenes. Thus, our dream images will naturally be surrealistic.

7. How telepathic communications work.

8. Why a person's aura would naturally contain a record of everything the person has gone through such that an individual who can see and read auras would be able to know a lot about the person by his or her aura.

9. How it is that our creator is a living and growing thing and thus can never be totally complete or absolutely perfect. This is why everything it creates could never be totally complete or absolutely perfect. And, why this would be a good thing and not a bad thing.

10. Why our spiritual advancements have fallen far behind our

amazing technical advancements. This is another way to see why we have not been getting much spiritual guidance for our behavior or much wisdom on how to live our lives more constructively and more meaningfully.

11. How artificial intelligence is able to learn things it was not designed or instructed to learn.

12. How quantum particles play a major role in enabling things to exist in our physical world and how space in our universe has to be a physical thing possibly made of quantum particles and how gravity could also be made by these same quantum particles.

13. UFOs and UAPs would appear as strangely shaped objects most likely because we can see them only partially and not completely.

14. Our mental abilities and memories reside with our spirit and not with our brain. Our brain is essentially the computer center for our body.

15. Things emerged in our universe by their spirits "thinking" them to emerge, and not by a Big Bang. That is why there is no antimatter yet to be found in our universe.

These are only a few of the examples of the attributes of life plausibly and logically explained by the new spiritual model. A large number of additional attributes are explained and presented in this book. For example, 59 attributes are listed in Chapter Two of this book. And more are presented elsewhere in this book.

Acknowledgements

The support and patience of my late wife Mae, my children Michael, Catherine and her husband Jon, my grandchildren Jake and Zack, my sister Mabel and her late husband Bob, my brother Walter and his wife Eleanor, my late sister Anna, my sister Pauline and her husband Bob, are greatly appreciated as I worked on this book and previously on References 1 and 2.

My best friend since childhood, the late Joe Yee, has been a wonderful influence on me my entire life. He was among the purest, most honest, and most caring people I know. Joe was the kind of person you could immediately become friends with the minute you meet.

I will always remember the numerous inspiring discussions regarding spirituality with the Rev. Tommy E. Smith, Jr. during the mid-1990 years that got me started writing my books. I also greatly value the stimulating and interesting discussions about spirituality from numerous points of view with Eric Johnson, the owner of the publishing company that published my books.

Always cherished is my relationship with my very special friend Rachel Burke whose ability to see and read auras has help me understand the meanings of the four unusual events described in Reference 1. The events took place shortly after my mother passed away, and they clearly indicated spiritual life continues after physical life is over.

The comments made by reviewing editor, DeAnne Musolf, regarding my first book have been a great source of encouragement. She commented that my book was a "Tour de Force" and that it was an honor to have worked on it.

I very much appreciated the continuing encouragements from Debra Holcomb of the California Academy of Sciences who understood how my work could benefit mankind and how it connects with the educational programs at the Academy for children and young people. The Academy's educational programs help children and young people gain the knowledge regarding how life works and would thereby help them become better adults.

Much acknowledgement goes to Ken Coit for the help he has given me and to many others as well and for being a great role model as a philanthropist.

Much appreciation goes to Alex Johnson for his excellent job designing this book.

My ever lasting love and appreciation go to my late parents Chester Q. and May W. Their hard working pioneering life and countless sacrifices gave my siblings and me the opportunity to achieve success in our lives.

Chapter One
Bridging the Gap between Science and Spirituality

1.1. Modeling Spirituality Along with Modeling Science Using Physical vs. Nonphysical Concepts

The need to take spirituality into consideration in science has been recognized and worked on by physicists for quite a while. They made a lot of progress over the centuries. A growing interest in the behavior and characteristics of quantum particles as they are associated with spirituality was also happening about the same time. Quantum particles are subatomic such as electrons, protons, neutrons, and any particles smaller.

When we get down to that level of existence, every physical thing is made of quantum particles. Even the existence of life seems to have something to do with the behavior and characteristics of quantum particles. Therefore, it is natural that any interest in how life works would include an interest in the behavior and characteristics of quantum particles.

A new and unconventional spiritual model recently formulated and presented in References 1 and extended in Reference 2 can explain how it is that quantum particles have a role in enabling the lives of living things to be possible in our physical world. This new spiritual model will be covered in greater detail as the presentation continues.

The behavior of quantum particles does not follow Newtonian physics. At times they appear as energy waves and at other times they appear as physical matter depending on the situation and on how they are observed. Thus, the theory of quantum mechanics, or quantum physics, was developed to describe the behavior and characteristics of quantum particles.

Quantum physics takes over where Newtonian physics ceases to apply. This is where elemental life becomes important to take into consideration in the modeling the behavior of physical things. The concept of elemental life as introduced in Reference 1 did not exist when quantum physics was developed. However, quantum physics captured the effects of elemental life on the behavior of physical things.

When we are dealing with large physical nonliving things, we can ignore elemental life since it played a negligible role in the behavior of what we would normally consider to be nonliving physical things. But, when we get down the quantum level of existence, we are dealing with behavior that is on the same scale as the effects of elemental life. Therefore, elemental life or at least the effects of elemental life needs to be taken into consideration.

Quantum physics could describe the behavior of nonliving physical things at the quantum level of existence to a large degree. But, quantum physics does not take elemental life completely into consideration, only most of the effects of elemental life. Therefore, it could not describe such phenomenon as quantum superposition and entanglement. The new spiritual model formulated in Reference 1 is a spiritual model that took science into consideration in its formulation. Thus, it was able to describe quantum superposition and entanglement as explained in detail in Reference 1.

Physics in general is a study of the behavior, characteristics, and workings of physical things in our physical world. Physicists have been able to describe physical properties and behavior of physical things by applying physical concepts to formulate scientific models that provide workable descriptions regarding how physical things work. The formulation of theories of physics in general has not explicitly take spirituality into consideration.

Life exists in our physical world; therefore, life is a part of our physical world. Therefore, from a physical point of view, science should be able to also explain how life works by applying physical concepts to formulate scientific models that could provide workable descriptions regarding how life works.

This means it should be possible to formulate scientific models that take spirituality into consideration by also applying physical concepts. One such model, perhaps the one most followed, was formulated by Lee Bladon as described in great detail in Reference 3.

Physicists apparently looked at several possible ways to develop scientific models that take spirituality into consideration. For example, they thought about whether reality is what it is because we observe it as being what it is or is it what it is regardless of whether or not we are observing it. Another example is they considered several possible ways consciousness could exist. These and other possible considerations are mentioned in References 3, 4, and 5.

Such considerations likely came from how a quantum particle would behave as an energy wave or as physical matter depending how it is observed, As mentioned earlier, when we get down to the tiny level of quantum particles, everything is made of quantum particles. Therefore, such considerations have a scientifically relevant reason to be explored when talking about modeling how life works by applying physical concepts.

But, it turns out science alone cannot explain everything about how life is possible by applying physical concepts. For example, religious models address some things that science could not explain such as how does consciousness come about and how DNA molecules could be formed? Religious models would simply say such things were created by God and would thus sidestep the need to take science into consideration.

Up to now, existing spiritual models such as religious models have not taken science into consideration. They rely instead on faith for their acceptance. By contrast, scientific models rely on logic, analytical methods, and verifiability through experiments and observations for their acceptance.

For this reason up to now a gap exists between scientific models and spiritual models such as religious models. This was explained by Kayume Baksh in Reference 4. Consequently, spiritual models that exist up to now and scientific models have not been able to connect together to form a unified description regarding how life works.

Nevertheless, in reality both scientific attributes and spiritual attributes are involved in enabling life to exist. Thus, both science and spirituality need to taken into consideration in formulating either scientific models or spiritual models in order to have models that can describe how life works.

This was made evident in the medical field in which how each vital organ was determined to work and how all vital organs are determined able to function in a coordinated manner to enable a living thing to be alive. In a sense it took scientific knowledge and spiritual considerations to "back engineer" how the creator must have applied scientific knowledge and spiritual knowledge to design and create living things.

This means ideally scientific models need to take spirituality into consideration and spiritual models need to take science into consideration in order to achieve a more complete, more accurate, and more unified description of how life works.

As mentioned by Lee Bladon in Reference 3, Einstein once said "Religion without science is blind, science without religion is lame", because "they are two sides of the same coin".

While much progress has been made by physicists to take spirituality into consideration in science, not much has been done by people working in the spirituality field to take science into consideration in spirituality. Thus, up to now the effort to bridge the gap has been pursued mainly by physicists. They came close to bridging the gap, but were unable to scientifically explain in terms of physical concepts how consciousness exists and how DNA molecules could be formed.

These two attributes are fundamental for explaining how life could exist and how life works. Therefore, to have anything close to resembling a unified description about how life works, physicists had to rely on the religious models to explain these two attributes in nonscientific terms since physicists were unable to model them in scientific terms.

But, now as mentioned earlier, a brand new unconventional spiritual model was recently formulated and presented in Reference 1 to provide **(1)** a new way of thinking about spirituality. The new spiritual model also introduced: **(2)** a set of new and unconventional nonphysical concepts for modeling the various attributes of spirituality and **(3)** the application of scientific and engineering logic as a way to take science into consideration in the formulation of the model. The model was **(4)** later extended in Reference 2 to include a possible origin of life by introducing a second set of unconventional nonphysical concepts for modeling the particular attributes of spirituality that is the origin of life.

As the presentation continues to where we are talking more extensively about this new spiritual model we will see **(5)** how the model could explain how it is that quantum particles could enable life to exist in our physical world. The new spiritual model could also explain: **(6)** how quantum particles would naturally sometimes appear as energy waves and sometimes as physical matter, **(7)** how consciousness could exist, and **(8)** how DNA molecules could be formed.

And, as described in detail in Reference 1 the new spiritual model could explain **(9)** how quantum superposition and entanglement could take place. The explanation is quite long and therefore not repeated here. Item **(9)** is aside from how the new spiritual model could explain Items **(1)**.through **(8).**

Apparently none of the topics in Items **(1)** through **(8)** could be explained by scientific models that took spirituality into consideration since none were mentioned in References 3, 4, and 5. These items seem to require a spiritual model that takes science into consideration to explain rather than a scientific model that takes spirituality into consideration to try to explain. This means in order to bridge the gap, an effort has to be made from the spirituality side of the gap instead of an effort strictly from the science side of the gap.

Why this is the case might be because in order for the creator to create living things, the creator would logically start by taking spirituality into consideration first and then take science into consideration second. In other words, the creator has to first think about the reason for creating living things before the creator would go about figuring out how to create living things.

Therefore, by analogy, in order to achieve a more complete and more accurate modeling regarding how life works, the formulation process needs to naturally and logically begin by taking spirituality into consideration first and take science into consideration second.

But, for the time being, let's briefly look at how physicists developed their models, using the model formulated by Lee Bladon in Reference 3 as an example.

The physicists were quite clever and creative in the way they applied physical concepts to describe the various nonphysical attributes of spirituality. Modeling anything nonphysical with physical concepts is bound to be somewhat complicated and requires a lot of cleverness and creativity. This is because there is an incompatibility problem to overcome when applying physical concepts to model nonphysical phenomena.

In Reference 3, numerous physical concepts were applied such as: **(1)** multiple dimensions beyond three dimensions, **(2)** five bodies postulated to make up a person, each embodying a different attribute of human nature, **(3)** multiple lifetimes a person would go through by means of reincarnation, **(4)** multiple universes, **(5)** morphic fields and various other kinds of energy fields, **(6)** seven planes or states of being a person could progressively achieve in a continuing quest for self improvement with each reincarnation, and **(7)** perfection being achieved upon reaching the seventh and highest plane, etc.

The various attributes of spirituality are discussed in the three books listed below, which are References 3, 4, and 5. General discussions are presented in References 4 and 5. An example of a detailed description of a scientific model that took spirituality into consideration was presented in Reference 3. The presentation in Reference 3 is quite extensive and complex and therefore was not repeated here.

- "The Science of Spirituality", by Lee Bladon, Lulu.com, August 1, 2007, and Made in the USA, Coppell, TX on May 2022, Reference 3.

- "Science and Spirituality, Bridging the Gap", by Kayume Baksh, Austin Macauley, November 30, 2021, Reference 4.

- "Spiritual Science", by Eric Dubay, Lulu.com, July 12, 2012, and Made in the USA, Las Vegas, NV in May 2022, Reference 5.

The physicists addressed almost every well-known nonphysical phenomenon associated with life such as: creation, evolution, reincarnation, sleep, dreams, out of body experiences, near death experiences, paranormal abilities, telepathy, remote viewing, premonitions, how creatures could find their way home from strange places, emotions, instincts, intuition, UFOs (unidentified flying objects) and UAPs (unidentified aerial phenomena), crop circles, spirits, souls, ghosts, God, other spiritual beings, and more.

A lot of these phenomena exist in the spiritual part of life instead of in the physical part of life. As the presentation continues it will become clear life consists of a spiritual part and a physical part. Thus, up to now, before the new spiritual model was formulated in Reference 1, a lot of such phenomena could not be explicitly and clearly explained as to how they work or how they exist.

But, from a scientific perspective, such things should be able to be scientifically explained in terms of physical concepts since they are all a part of life and life is a part of our physical world. This is a very logical way to look at it, and the physicists were able to developed scientific models that can explain such things by very cleverly and creatively applying physical concepts.

For example, a person has numerous nonphysical attributes such as major senses, emotions, mental abilities, awareness, etc. Therefore, from a scientific perspective a person could be perceived as consisting of five different bodies: **(1)** a physical body, **(2)** a body with senses that are tuned to our physical world, **(3)** a body with emotions, **(4)** a body with mental abilities, and **(5)** a body with awareness. Bodies **(2)** through **(5)** are nonphysical. But, the concept of bodies is a physical concept. Thus, this physical concept was applied cleverly and creatively to describe four nonphysical spiritual attributes.

Another example in Reference 3 depicts a person as constantly striving to become better and to ultimately become perfect. A person would thus be depicted as working toward being better during each lifetime as he or she goes through a sequence of reincarnations. The person would thus progress upward through seven planes of existence, one plane at a time with each reincarnation.

The person would be physical and reside in our physical world in the first three planes. The person would become a spiritual being in the fourth through seventh planes and would thus no longer reside in our physical world but would reside instead in presumably the spirit world. Perfection is achieved upon reaching the seventh and highest plane. There, the person would become another God who could create universes. Again, planes are physical concepts being applied to describe nonphysical attributes of spirituality, attributes that are associated with humans striving to become better.

I suspect the idea of humans eventually becoming gods and can thus create universes was included in the physicists' model to possibly explain how multiple universes could be possible.

An interesting observation is that if God were perfect, then God needs nothing more to improve. I suspect the primary source for this idea would be the various religious models in which God was perceived to be all knowing and all powerful, and the secondary source would be our universe.

Our universe is perceived as having nothing more to improve. Every physical phenomenon that could take place in our universe is already taking place, and everything is working just right to enable life to exist in our universe. Therefore, our universe needs nothing more to take place and thus could be perceived as being perfect. This means since God is perceived to be the creator of our universe and our universe is perceived to be perfect, this means God has to be perfect in order to create something that is perfect.

But, a problem exists with perceiving the creator, God, as being perfect. It leaves no room for further improvement and therefore not room for further growth. Therefore, if life were perfect, than life would have no room for improvement and thus no room for a purpose except perhaps to eventually create more universes. It is thus interesting what is said in Reference 4.

As pointed out by Kayume Baksh, Reference 4, it is not clear if there is a predetermined purpose for humans to exist on Earth. Paraphrasing what was said, "Thus, it is essentially up to each individual to find personal purpose in his or her life." This goes along with how we humans keep pursuing knowledge about how things work in our universe without addressing why our universe was created in the first place or what is the purpose we humans have for being created to reside on Earth.

However, as the presentation continues, we will see how it is that the creator is bound to have a reason for creating our universe and thus we humans are bound to have the responsibility for helping our universe fulfill its reason for being created. Therefore, we human must have a purpose for being here on Earth. It is just that up to now we have not been able to identify this purpose. However, this has now changed with the new spiritual model having been recently formulated in Reference 1.

We will see in the next section of this chapter how science was taken into consideration in the formulation of the new spiritual model in a natural way as presented in Reference 1.and extended in Reference 2. The new spiritual model is able to reveal the reason our universe was created and the purpose and responsibility we humans have for having been created to reside on Earth.

The new spiritual model was formulated with a new way of thinking regarding spirituality and by applying unconventional nonphysical concepts instead of conventional physical concepts to model the attributes of spirituality. Spirituality is nonphysical, therefore applying nonphysical concepts would naturally be more compatible with it and would make the modeling process more complete and more accurate.

Thus, the new spiritual model was capable of explaining the attributes of spirituality more simply, more completely, and more accurately. This includes being able to explain the existence of consciousness and how DNA molecules could be formed. The physicists' model was unable to explain these two attributes.

Other than that, the new approach and the physicists' approach were both found to be fairly successful.

As explained in Reference 1, the new spiritual model achieved a high level of confirmation by being able to explain just about any common everyday experience and observation we can think of. Thus, it is easy for the new spiritual model to be a natural part of everyday life and for us to naturally stay tuned into its spiritual guidance every moment of every day. This is how a spiritual model should be.

The model also clearly shows how the creator is a living and growing thing and thus always has more room to grow and improve. This means the creator is never complete or perfect since there is always more that it could become.

Therefore, the model was able to show how it is that our universe has a reason to be created, and that we humans are here on Earth to help our universe fulfill the reason it was created. This is a major difference in findings between this new approach and the physicists' approach and also between the new approach and the conventional approach taken by conventional spiritual models such as the various religious models that were available up to now.

As the presentation continues, the potential positive outcome from having scientific models that take spirituality into consideration and spiritual models that science into consideration is that mankind will hopefully eventually become more self inspired and more self motivated to behave overall better and wiser. This could also hopefully eventually lead to mankind no longer make wars, mistreat one another, and do criminal acts of all kinds.

This is the hope for writing this book and References 1 and 2 and for also emphasizing the achievements made by physicists regarding their modeling of how life works. It might take hundreds or thousands of years to achieve, if ever. Still, we need to try, and we need to start somewhere.

1.2. Modeling Science Along with Modeling Spirituality Using Scientific and Engineering Logic

As indicated earlier, in order to create things as complex as our universe and us humans and other living things, the creator has to apply both spiritual knowledge and scientific knowledge. This means, for us humans to formulate models that can describe more completely and more accurately how life works, both spirituality and science need to be taken into consideration.

Until fairly recently, we formulated models by applying science only or by applying spirituality only. Thus, as explained in Reference 4, a gap exists between scientific models and spiritual models. This means a model that can describe how life works and that can be verified by common everyday experiments and observations has not been developed until fairly recently.

What we needed are scientific models that take spirituality into consideration and spiritual models that take science into consideration.

Physicists were the first to try bridging the gap by formulating scientific models that took spirituality into consideration. Several approaches were explored as explained in References 3, 4, and 5. The model developed by Lee Bladon in Reference 3 was used here as an example of what the physicists have achieved.

The new spiritual model recently formulated in Reference 1 and extended in Reference 2 is likely to be the first spiritual model to take science into consideration. Scientific and engineering logic was applied throughout the formulation process. We will see how this was done as the presentation continues. A more detailed description of the formulation process is presented in References 1 and 2.

The formulation process introduced a new way of thinking about spirituality. It also introduced several new unconventional non-physical concepts for describing the various attributes of spirituality and also how things work in the spirit world.

For example, the new spiritual model explains in a logical manner how every living thing in our physical world has a spirit in the spirit world that enables the physical body of the living thing to exist and be a living thing in our physical world. Thus, a living thing consists of two parts, a spirit in the spirit world and a body in our physical world.

Accordingly, life consists two parts, a spiritual part going on in the spirit world and a physical part going on in our physical world, and both parts going on at the same time. The spiritual part enables the physical part to exist. The spiritual part would logically go on forever in the spirit world whereas the physical part would last only for a short while in our physical world.

This is different from how conventional religious models seem to imply spiritual life begins after physical life ends.

The new spiritual model is also able to explain in a logical manner how consciousness exists and how DNA molecules could be formed. The religious models do not address these two important attributes about how life works. If they were to address them, the religious models would likely simply say these attributes were created by God, and we would have to accept them based on faith instead on logical reasoning.

It also turns out that the new spiritual model is simpler, more complete, and more accurate than are the physicists' models mainly because nonphysical concepts were applied to model the nonphysical attributed of spirituality instead of apply physical concepts to do so as the physicists have done. Nonphysical concepts are more compatible with modeling nonphysical spiritual attributes than physical concepts could be.

Nevertheless, the approach taken by physicists to formulate their models and the approach taken to formulate the new spiritual model are both fairly successful in their ability to describe how life works.

A difference is the new spiritual model was able to explain in a logical manner the following additional attributes about how life works that the physicists' model did not explain: **(1)** how past universes have to have been created before our universe was created, and more universes are bound to be created in the future, **(2)** how allergies might have developed among some living things on Earth, **(3)** how oneness naturally forms everywhere within the spirit world, **(4)** how every person's spirit is sharing a major part of itself with every other person's spirit in the spirit world; thus, we are all a major part of each other in the spiritual part of our lives, **(5)** how our consciousness, intelligence, mental abilities, memory, emotions, instincts, intuition, etc. all reside with our spirit in the spirit world and not with our brain in our physical world as commonly assumed, **(6)** how sleep enables our spirit to restore our body from the wear and tear it gets during waking hours, and **(7)** why dreams tend to be surrealistic.

And, again, as mentioned earlier, the new spiritual model is able to explain: **(8)** how quantum particles have an important role in enabling physical things to exist in our physical world, **(9)** how it is that quantum particles could appear as energy wave or as physical matter depending on how they are observed, **(10)** how consciousness can exist, and **(11)** how DNA molecules could be formed. Much more are described as the presentation continues.

Also as mentioned earlier and explained in detail in Reference 1, the new spiritual model is able to **(12)** provide a plausible spiritual explanation for how quantum superposition and entanglement can occur. This is a phenomenon that physicists have not been able to explain. Einstein once described it as "spooky action at a distance".

The formulation process for the new spiritual model began with the logical and indisputable statement that **"something, somewhere, somehow knows how to enable our universe and**

everything in it to exist." This is the first stage at which scientific and engineering logic was applied in the formulation process. This logic was applied throughout the formulation process as will be shown as the presentation continues.

The formulation process also introduced the new unconventional nonphysical concept **"pieces of knowledge"** to describe how things work in the spirit world. How this was done is described in detail in Reference 1.

Knowledge obviously comes in pieces. This is why education could be achieved stepwise going form first grade, to second grade, to third grade, etc. It is also why any specific subject could be taught with a series of lessons, with each lesson covering a certain portion of the total knowledge making up that specific subject. It is also why each lesson can build upon what was covered in preceding lessons.

Most importantly, it is why knowledge learned in one field could also be applied in other fields such that technical advancements would be sped up and enhanced by such interactions among numerous fields. Examples are **(1)** biological knowledge applied in the medical field, **(2)** mathematic, physics applied in the development of high technology, and **(3)** material science applied to make space exploration possible. Etc.

Thus, the nature of knowledge is that it comes in pieces.

Another new unconventional nonphysical concept was introduced in Reference 2 to describe a possible origin of life as being formed by the four spiritual qualities; "consciousness, intelligence, curiosity, and wisdom". How this works is explained in detail in Reference 2 and also briefly as the presentation continues.

As mentioned earlier, the new spiritual model achieved a high level of confirmation by being capable of providing logical spiritual explanation for just about any common everyday experience and observation we can think of. This is described in detail in References 1 and 2. Additional and more rigorous confirmations are described in this book as the presentation continues.

Some good matches were found between the new spiritual model and the scientific model in Reference 3. This lends confirmation for both models. Examples of the match are presented later as the presentation continues.

A few important differences between the new spiritual model and the scientific model were also found, and these also explained as the presentation continues. Some of the differences came from the difference in what the physicists wanted to accomplish and what the new spiritual model was formulated to accomplish.

The physicists were interested in understanding how life works in our physical world. The new spiritual model was formulated with that interested in mind as well but also with how the new spiritual model could inspire and motivate people to behave overall better and wiser

Such differences are aside from the fact that the physicists were developing their models from the science side of the gap while the new spiritual model was formulated from the spirituality side. Up to now the main obstacle to bridging the gap has been a lack of progress being made from the spirituality side of the gap. But, now major progress has been achieved as represented by the new spiritual model.

In summary, the differences between what the physicists were trying to accomplish and what the spiritual model was formulated to accomplish are as follows:

- **What the Scientific Model Formulated by Physicists in Reference 3 Was to Accomplish.** As mentioned earlier, physicists have been interested in how life works for quite some time. Accordingly, they developed scientific models that took spirituality into consideration to increase this understanding. A good example of such models is formulated by Lee Bladon in Reference 3.

  Physicists realized both scientific knowledge and spiritual knowledge have to be a part of such models. Physicists are used to applying physical concepts to develop scientific models. Therefore, they would understandably explore various approaches to model how life works by also applying physical concepts. This might have been also influenced by how the traditional religious models were formulated by applying physical concepts. Examples of physical concepts applied included places such as heaven, hell, and a kingdom of God.

  The physicists explored various approaches. One approach has to do with whether reality is as we observe it because we are observing it, or is it as it is regardless whether or not we are observing it. This might have stemmed from how a typical quantum particle would behave sometimes as an energy wave and sometimes as physical matter depending on how it is being observed. When we are at that tiny level of existence, every physical thing is made of quantum particles. Thus, the behavior and characteristics of quantum particles have real meaning from a scientific standpoint when developing models that describe how life works.

  Another approach the physicists explored has to do with what is consciousness and how it could exist. Does it exist only with things we consider to be living things or does it exist with everything whether it is considered living or nonliving?

How they would go about developing their models would depend on which of many possible approaches they would choose to follow.

In the model presented in Reference 3, the physicists achieved significant success in developing a workable model by cleverly and creatively applying physical concepts to describe nonphysical spiritual attributes. It was done in a manner that is realistic, logical, analytical, nonjudgmental, and in some instances verifiable with real life experiences and observations.

This is the model that is chosen in this book to represent what the physicists have accomplished starting from the science side of the gap.

- **What the New Spiritual Model Formulated and Presented in References 1 and 2 Was to Accomplish.** Starting from the spirituality side of the gap, a new spiritual model was formulated in a manner that is logical, analytical, nonjudgmental, and verifiable with real life experiences and observations.

    The new spiritual model was to accomplish the following:

    1. The new spiritual model turns out to be more complete and more accurate than are the religious models in terms of modeling how life works. It essentially picked up where the religious models left off. Thus, the new spiritual model acknowledges the accomplishments made by the religious models and it builds upon what the religious models have achieved.

    2. Accordingly, the new model was not meant to replace the religious models. It complements them instead. Later as the presentation continues a stepwise progression process was

suggested to help mankind become more knowledgeable about how life works. The religious models would constitute the first step.

3. The hope is that becoming more knowledgeable about how life works would help mankind become more self inspired and more self motivated to behave overall better and wiser. For example, mankind would stop making wars, mistreating one another, and doing criminal acts of all kinds. Thus, mankind could then use its time and energy instead to form the wisdom needed to figure out how to live life more constructively and more meaningfully and to actually live life in this manner.

Existing faith based religious models must have been the best that could be formulated at the time and would be based on the knowledge the people possess at the time. The religious models must have also been quite in sync with the way of life was during that period.

People back then were likely not as analytical and technically oriented as they are today. Thus, a reliance of faith for acceptance might have been more common and more acceptable at that time. Therefore, the religious models were bound to be more effective in inspiring people to behave overall better at that time than the religious models are today. The way of life has changed a lot with time.

But, regardless, the models were unable to make much progress inspiring mankind to behave overall better in the following thousands of years. People continued to make wars, mistreat one another, and do criminal acts of all kinds the whole time.

The religious models were not even able to stop some Catholic priests from mistreating various people for decades leading up to even the present. The worse examples are their mistreatment of indigenous people in extremely terrible ways in the 1800's and 1900's in United States and Canada. They took children away from their parents and put them in residential schools where hundreds died from being horridly abused and were then buried in mass graves, References 6 and 7. Other reports about the residential schools indicated children were forced to dig graves for fellow students that died.

This is clear evidence of the significant incompleteness and inaccuracy of the religious models. Again, this does not mean the religious models were incorrect. They are correct, but they are also incomplete and inaccurate, perhaps very much incomplete and inaccurate in certain respects. But, we need to realize nothing is ever totally complete or absolutely accurate.

As the presentation continues we will be able see how this is the case because the creator of everything could never be totally complete or absolutely perfect. This is in spite of how we might perceive God to be all knowing and all powerful. Such perceptions apply only in comparison with the knowledge mankind possesses.

In order for the creator to be a living thing, it has to also be a growing thing. This means there will always be more that the creator could become as it continues to grow. Thus, at any given point in time the creator could not be totally complete or absolutely perfect since there is always more that it could become as it continues to grow.

If the creator is imperfect then everything it creates would be imperfect, and we can see how this is the case for everything that exists on Earth. Conversely, everything on Earth being imperfect could indicate the creator must be imperfect in the first place.

Frankly, it is a good thing everything is imperfect. Otherwise everything of the same kind would be identical, and that would be boring and problematic. For example, we wouldn't be able to tell which person is the one we married or which car is ours. We would be bored to death because innovations and inventions could not take place.

If we insist our creator is already perfect, then we immediately put a barrier to our pursuit of spiritual advancement by hindering any incentive to advance further spiritually. We wouldn't even bother to figure our why our universe was created in first place.

Life consists of a spiritual part and a physical part such that true advancement would consist of both spiritual advancement and physical (technical) advancement. We humans tend to think we are making significant advancements by simply making significant technical advancements without making much spiritual advancements to go along.

Mankind made amazing technical advancement but not much spiritual advancement. Thus, we apply our technical advancements in destructive and meaningless ways as much as in constructive and meaningful ways. This is because we have not achieved the spiritual advancements needed to guide us how to carry out our lives more constructively and more meaningfully.

This imbalance in the two parts of our advancements and in our knowledge about how life works is preventing us from being able to form the wisdom we need to figure out how to live our lives more constructively and more meaningfully.

The religious models are also gradually becoming increasingly out of sync with the population, because the way of life keeps changing with time. Consequently, the effectiveness the religious models originally had has gradually declined with time.

For example, today, church attendance is declining, and people are increasingly exploring other means and other criteria in their search for their personal spirituality. A reliance on faith for acceptance is less likely to resonate with today's population with their broader range of knowledge and their more technically and analytically oriented way of life.

Again, all that has happened does not mean the religious models are incorrect. The religious models are correct but incomplete and inaccurate. More specifically,

1. Religious models are correct in terms of their spiritual guidance.

2. But, they are incomplete and inaccurate in terms of not having logically based models to backup what they are teaching. Thus, religious models must rely on faith for their acceptance.

3. Time has shown how faith is unable to inspire people enough to have people behave overall better and wiser even over thousands of years because of the models' incompleteness and inaccuracy.

4. In addition, the religious models' reliance on faith for their acceptance, and the stories the models tell, are becoming increasingly out of sync with today's population.

Reliance on faith for acceptance is OK as long as logically based models exist to back it up. We work with such situations all the time in today's way of life. For example, we follow building codes by relying on faith that the codes will work. This is OK because we know the codes are backed up with logically based structural analytical models that have been verified with real life experiences and observations.

Thus, a conclusion is that something spiritual has been missing in the way we humans are living our lives, because our purpose for being here on Earth could not possibly be to behave as overall badly as we have been behaving. We need something that could help people understand the reason our universe was created and that we humans have the responsibility for helping our universe fulfill its reason for having been created.

What the physicists accomplished is realistic, logical, analytical, nonjudgmental, and verifiable. But, it does not reveal the reason our universe was created or that mankind has a purpose to fulfill for being here on Earth.

We need to know more completely and more accurately both the scientific knowledge and the spiritual knowledge about how life works in order to form the wisdom needed to figure out how to carry out our lives more constructively and more meaningfully.

> **1. Scientific knowledge helps mankind understand how things work in our physical world so that mankind could become proficient at doing things in our physical world.**
>
> **2. Spiritual knowledge helps mankind understand how things work in the spirit world so that mankind could become more self inspired and more self motivated to carry out our lives in a manner that is more constructive and more meaningful while we are alive in our physical world.**

Thus, it is clear that we need both science and spirituality to provide mankind with the knowledge to form the wisdom needed to be able to figure out why our universe was created and what our purpose is for being here on Earth. Then mankind would be more self inspired and more self motivated to behave in a manner that is more constructive and more meaningful.

We have gotten a lot of the scientific knowledge we need from science, but up to now we have not been getting much of the spiritual knowledge we need from spirituality. The new spiritual model has the capability to provide us the spiritual knowledge we need.

Its formulation process explained in References 1 and 2 is quite complex and long, and therefore was not repeated here. However, a briefly explanation of it and the main features of the new spiritual model are summarized as the presentation continues.

Chapter Two

The Formulation of an Unconventional New Spiritual Model Introduced A New Way of Thinking and New Unconventional Nonphysical Concepts

2.1. The New Spiritual Model Has to Be Applicable Throughout Our Universe

For convenience, most of the features of the new spiritual model formulated in References 1 and 2 and the information it revealed so far about the spirit world and how life works are collected into this chapter. The few features already mentioned earlier in this book are also included.

Many of the features and the information were revealed by the new spiritual model after References 1 and 2 were published. Therefore, a lot could be found in this book that is not found in References 1 and 2. It is also quite likely the model will reveal additional information about how life works after this book is published.

Because the new spiritual model was able to provide plausible logical spiritual explanations for just about any everyday experience and observations we can think of, as explained in Reference 1, the new spiritual model has proven to be fairly complete and fairly accurate in its modeling of how things work in the spirit world and thus how life works.

However, nothing could ever be totally complete or absolutely accurate. This is because, as explained in Reference 1, the creator of everything (the spirit world) could never be totally complete or absolutely accurate. Therefore, nothing it creates could ever be totally complete or absolutely accurate either. This includes the new spiritual model.

Every piece of information about how things work in the spirit world and how life works as revealed by the new spiritual model is consistent with real life experiences and/or observations. Therefore, even if direct confirmation of some of such information is not possible, every piece of information is either currently real or has a high probability of having been real in the past.

For example, the new spiritual model indicates multiple sequences of universes were likely created in the past and that the process of creating them is still ongoing. Our universe was created as part of creating one of the sequences. This means many universes were brought into being before our universe was and more will continue to be brought into being.

This cannot be directly confirmed, but it is consistent with how things work in the spirit world and how life works. Therefore, this information is highly likely to be real. This and numerous other pieces of information are topics of discussion as the presentation continues.

The following requirements were imposed during the formulation of the new spiritual model:

1. The same creator created our entire universes and not just Earth. Therefore, the new spiritual model has to be applicable throughout our universe and not just on Earth.

2. The formulation process for the new spiritual model must take science into consideration.

3. The new spiritual model has to be nonjudgmental, and it must rely on logic for its acceptance and not on faith. A reliance on faith is not resonating well with today's more educated and more analytical thinking population.

4. The model has to feel comfortably and naturally as being a part of everyday life such that we could naturally stay in touch with it and its spiritual guidance every moment of everyday. This is how spiritual models should be. After all, spirituality is a natural part of life as explained by the fact that life consists of a spiritual part and a physical part.

To meet these requirements, the formulation process started from scratch instead of from any currently existing spiritual model. Accordingly, the formulation process began with the logical and indisputable statement that **"something, somewhere, somehow knows how to enable our universe and everything in it to exist."**

The rest of the formulation process is described in detail in References 1 and 2. The new spiritual model ended up meeting all four requirements as indicated in References 1 and 2. The formulation process was quite long and therefore was not repeated here. However, its main points are described as the presentation continues.

Up to now, we humans on Earth have not thought about the possibility that the perceptions embedded in our current conventional spiritual models are not likely to be accepted by living things residing elsewhere in our universe. Therefore, our current conventional spiritual models are not likely to be applicable throughout our universe.

Many alternative perceptions are possible for formulating spiritual models even on Earth as indicated by how a lot people are now searching for their personal spirituality by following criteria that are different from those of our current conventional spiritual models. It is also indicated by the thousands of books written on numerous topics associated with spirituality that are not a part of our current conventional spiritual models.

This means models such as our current religious models are not likely to be followed elsewhere in our universe, and therefore they would not meet Requirement number 1. This is one of many indications that while our current religious models are correct and good they are also quite incomplete and inaccurate. Therefore, their spiritual guidance is also correct and good but is quite incomplete and inaccurate.

This could explain why our current religious models have not been able to motivate humans to stop making wars, mistreating one another, and doing criminal acts of all kinds. They are simply not complete enough and accurate enough to be effective enough.

2.2. Helping the Spirit World Learn Ways Living Things Can Be Physically Separate While Also Be Spiritually a Part of Each Other by Us Humans Learning to Do So

The new spiritual model enabled several observations to be made regarding the spiritual oneness and various other attributes that naturally and automatically develop in the spirit world. The observations are listed below. Unless you are familiar with References 1 and/or 2, many of the terms, concepts, and attributes stated are going to be unfamiliar at this stage of the presentation. However, they will be explained later in this chapter as the presentation continues.

The reason these observations are presented before the terms, concepts, and attributes are individually explained is because the terms would often become clear as they are used in context. Another reason is by presenting the observations first would make the individual explanations that come later easier to follow as to the roles the terms play in describing how things work in the spirit world and how life works in general.

1. **The Concept of Opponent:** Because of the spiritual oneness in the spirit world, the concept of opponent is only a concept and not a practice in the spirit world. Every spiritual entity is directly or indirectly a part of every other spiritual entity in the spirit world. Therefore, battles, mistreatments of one another, and criminal acts of any kind are not going to happen in the spirit world. Instead, working together to resolve issues is natural.

The way things work in the spirit world is very different from the way things work in our physical world, our universe. Every thing in our physical world is a separate entity such that physical oneness does not exist, and the concept of opponent could become a practice as it is very much a practice on Earth.

Our universe was created to be physical and with entities that are separate from each other for a reason as we will see as the presentation continues.

2. **Oneness and Separateness:** Because of the spiritual oneness that pervades the spirit world, it is natural the spirit world would, on its own, generate a lot of new pieces of knowledge by spiritual entities that are directly or indirectly a part of each other. Thus, an excess of new pieces of knowledge that pertain to spiritual entities that are directly or indirectly a part of each other would periodically result. Thus, an imbalance in the state of knowledge in the spirit world would periodically develop.

This imbalance needs to be restored to reasonable balance with new pieces of knowledge that pertain to entities that are separate from one another.

3. **Nonphysical and Physical:** Another kind of imbalance that periodically occurs in the spirit world on its own is an excess of new pieces of knowledge that pertain to nonphysical spiritual entities because all spiritual entities in the spirit world are nonphysical.

This imbalance needs to be restored to reasonable balance with new pieces of knowledge that pertain to entities that are physical.

4. **Issues Are Forms of Imbalances:** In addition to issues that develop on their own in the spirit world, any issue that develops in any universe would have a spiritual part existing in the spirit world. Thus, the spirit world will have all kinds of issues and therefore all kinds of imbalances at just about any point in time.

For example, we humans face numerous issues that develop in our lives everyday on Earth. Examples are financial issues, children growing up and wanting more independence and freedom, equipment wearing out and needing replacements, problems caused by global warming, one or more conflicts of various kinds are always happening on one or more locations on Earth, etc.

Each such issue that developed on Earth will have its spiritual part in the spirit world, and each will thus be an imbalance that developed in the state of knowledge of the spirit world.

5. **New Universes Are Created for a Reason:** As explained in Reference 1, the state of knowledge in the spirit world has to be kept reasonably balanced in order for the spirit world to be able

to form the wisdom required to figure out how to keep the spirit world itself viable. Thus, when an imbalance of any kind develops, the spirit world would design and create a new universe that could generate the kinds of new pieces of knowledge that would help restore reasonable balance for that particular imbalance.

A large number of imbalances are bound to exist at any given point in time. Thus, the spirit world would typically have to design and create multiple new universes to handle all the imbalances that need rebalancing. A single universe could handle more than one imbalance. How many a single universe could handle would partly depend on how well and how many of the imbalances could be handled together by a single universe.

Based on the discussions in Items 2 and 3 above, new universes would commonly be created to have entities that are separate from each other and/or that are physical, Our particular universe was created to handle these two imbalances together along with numerous other imbalances such as those mentioned in Item 4 above.

Based on the discussion in Item 4 above, new universes would have various other attributes depending on the various other kinds of issues that developed on already existing universes that could not be adequately handled by those already existing universes. For example, global warming could be an issue that developed on Earth that could end up not being adequately handled on Earth.

Each new universe would be different and unique as explained in detail in Reference 1. Differences could include each having a different dimensionality, being made of different kinds of substances and energies, having forces other than gravity if any, being all liquid or all gaseous or all solid, etc.

6. **Being Spiritual and Physical at the Same Time:** Since our universe was created such that entities are physical and are also separate from each other, it is one of those universes in which part of their purposes is to help restore reasonable balance for the two periodically occurring imbalances described in Items 2 and 3 above.

Our universe could be created to also handle other issues as well such as those mentioned in Item 4 above. Some of those in Item 4 might be new ones that developed on Earth, and if Earth could not adequately handle them they could become issues to be handled by new universes that will be created later.

Based on the discussions so far, part of the reason we humans are here on Earth is to accomplish the following: **We humans are here on Earth to help the spirit world, which is the creator of everything, learn ways to carry out life constructively and meaningfully by living entities that are physically separate from each other and that are at the same time spiritually a part of each other.**

We might ask: How could we humans be both physically separate and spiritually a part of each other at the same time? The answer is everything in our physical universe, including us humans, has a spirit which is a spiritual entity in the spirit world enabling the physical part of it to exist in our universe. Thus, every physical thing on Earth has two parts, a spiritual part in the spirit world where it is a part of everything else in the spirit world and a physical part in our universe where it is separate from everything else.

Thus, when we are carrying out our lives we are doing it as

entities that are physically separate from each other that are also spiritually a part of each other. We are to learn how to do carry out our lives constructively and meaningfully. Since our spirits are a part of the spirit world, what our spirit learns would become a part of the spirit world and thus it would also be what the spirit world learns.

Therefore, when we learn something, we are really helping the spirit world learn that something. But to be helpful and beneficial for the spirit world, whatever it is that we learned has to be learned in a manner that matches how things work in the spirit world in order for the spirit world to be able to use it most effectively.

In other words, our behavior on Earth has to be compatible with how thing work in the spirit world in order to be most beneficial for the spirit world, and for ourselves since our spirit is a part of the spirit world. The topic of compatibility with the spirit world is covered in the next item.

7. The Need to Be Compatible With the Spirit World: To "carry out life on Earth as entities that are spiritually part of one another" is part of carrying out our life on Earth in a manner that is compatible with how things work in the spirit world.

This compatibility is important because in order for the newly generated pieces of knowledge to counteract the existing pieces of knowledge that are causing an imbalance, the new pieces of knowledge have to be generated in a manner that matches how things work in the spirit world.

By analogy, it like if the transmission of a car is low in transmission fluid, we need to add the kind of fluid that is compatible with how the transmission works. We cannot add just any kind of fluid. Any other kind of fluid would not be effective in helping the transmission to work. In fact the wrong kind of fluid could make the transmission work badly.

> a. Because oneness naturally and automatically forms within the spirit world, the way things work in the spirit world would only be constructive and never be destructive, and the things that are constructed would always be meaningful and never meaningless.

> b. Because every living thing has a spirit that resides in the spirit world and a physical part that resides in our physical world, the physical part has to behave in a manner that is constructive and meaningful in order to help restore reasonable balance in the spirit world.

When this happens, it means the spirit is piloting the physical part a manner that matches how things work in the spirit world. This is necessary in order for the spirit's behavior to be effectively in accomplishing things in the spirit world that would be constructive and meaningful.

In this case, it is to help restore reasonable balance in the state of knowledge in the spirit world.

8. We Have a Partnership Relationship with the Spirit World: One way to interpret what is presented so far is each of us has a partnership relationship with the spirit world. This is because our spirit is a portion of the spirit world, and it was selected by the spirit world to help restore reasonable balance in the spirit world. Thus, a partnership is naturally formed between the spirit world and our spirit.

Our spirit then enables the physical part of us to be formed and exist in our physical world. Thus, the physical part of us is residing in our physical world as a representative of our spirit.

As with any partnership, our partnership with the spirit world is intended to be mutually beneficial for both the spirit world and us. This also means we need to recognize we have the responsibility to make the partnership work as intended.

Thus, we are here to care for the spirit world as well as receive care from the spirit world. So far, we humans on Earth have not been caring for the spirit world very well by making wars, mistreating one another, and doing criminal acts of all kinds. We have not taken up our responsibility for making the partnership work properly.

Looking at it as a partnership would give us more incentive to want to behave better and wiser and thus do things more constructively and more meaningfully in a longer term and longer range basis.

9. **Being Spiritually Generous:** In my mind, it is important for us humans on Earth to be spiritually generous while we are working on living life as entities that are physically separate from each other while at the same time being spiritually a part of each other.

We know being physically generous means being generous in giving physical things including money. We would give such things simply because we want to without expecting anything in return.

Being spiritually generous would work the same way. The difference is it means being generous in giving spiritual things such as care, empathy, sympathy, concern, support, devotion, love, acceptance, inclusion, a friendship attitude, a partnership attitude, appreciation, lifting up someone's spirit, etc. We would give such things simply because we want to without expecting anything in return.

10. **Diversity, a Vital Attribute of the Spiritual Part of Life:** In my mind, another important part of carrying out life constructively and meaningfully while we are working on living life as entities that are physically separate from each other while at the same time being spiritually a part of each other is to be knowledgeable about the following:

> a. Diversity in all its possible forms is a natural and valuable part of the spirit world's ability to maintain its own viability. The new spiritual model is able to show how this is the case as explained in References 1 and 2.

> b. Further discussion regarding the enormous value of diversity is presented as the presentation continues. Basically, every part of the spirit world is as valuable as any other part. Thus, every spirit is as valuable as every other spirit. The spirit world needs all of itself to be participating in keeping the spirit world viable. Every part of itself and every spirit in the spirit world have a role to play in the effort.

> c. Both the spiritual part of life and the physical part of life are going on at the same time when we are alive on Earth. The spiritual part is way more important than the physical part. This is because the spiritual part of a living thing is really what the life of a living thing is, and it is the part that lasts forever. The physical part is only a temporary representation of the spiritual part and it is for fulfilling a purpose. Thus, it lasts only for a while.

These attributes and how they work are described and explained by the new spiritual model.

We humans on Earth need to embrace and value diversity in all its forms. As will be covered later as the presentation continues, diversity in all its forms is in a sense a natural gift from the spirit world to all highly intelligent living things in any universe to enhance their ability to form wisdom regarding how to living their lives constructively and meaningfully.

This because diversity widens the range of knowledge available to us regarding how life is working and therefore enhances our ability to form wisdom that enables us to be more able to figure out how to live our lives more constructively and more meaningfully.

Realizing this takes a certain amount of wisdom to begin with. Thus, it takes wisdom to enable us to figure out how to form more wisdom. This suggests that one thing we humans could do is to "start this ball rolling" by helping humans as a whole to gain enough wisdom in the first place, enough for humans to then figure out how to form more wisdom.

11. Needing to Be More Convinced the Spiritual Part of Life Is Real: For us humans on Earth, destructive and meaningless acts such as wars, mistreatments of one another, and criminal acts of all kinds have become practices.

We seem to be not fully convinced that the spiritual part of life is real, is going on the same time the physical part of life is going on, and is way more important than the physical part. We seem also not fully convinced that the spiritual part would go on forever if it exists.

I think part of the reason we are skeptical is our current conventional spiritual models are not complete enough and not accurate enough in describing the spiritual part of life, and acceptance of the models is based on faith instead of based on logic.

The new spiritual model does a much better job in both of these areas and therefore would do a better job of convincing people that the spiritual part of life is real and would last forever

12. **An Egg Transforming into a Chick Is an Example of How the Spiritual Part of Life Is Real:** An observation we can make every morning if we eat eggs for breakfast. The observation would clearly indicate the spiritual part of life is real as follows:

> a. If we crack open an egg and look at the contents contained in the egg, we could not tell which part of the contents would form the wings of the chick or the legs of the chick or the head of the chick, etc. However, something somewhere somehow knows how to transform an egg into a chick. That something has to be the spiritual part of life.

> b. The chick would immediately know how to walk, how to eat, be conscious of things around it, etc. Were all such knowledge, intelligence, consciousness, and abilities embodied in the contents of the egg? If not, where did they all come from? They have to come from the spiritual part of life.

> c. In addition, that spiritual part of life has to exist and functioning in the spirit world the same time the egg, the physical part of life, is existing in our physical world. The spiritual part of life does not begin after the physical part of life ends as we are often led to believe by some of our existing conventional spiritual models.

13. **We Are All a Part of Each Other in the Spiritual Part of Life:** All spirits are spiritual entities, and all spiritual entities are directly or indirectly a part of each other. All human spirits are more than 90% a part of each other in the spiritual part of life as indicated by how our genes are more than 90% a part of every other person in the physical part of life.

Thus, while we are all physically separate from each other in the physical part of life, our spirits are all more than 90% a part of each other in the spiritual part of life.

The spiritual part of life is more important than the physical part of life. Our spirits are who we really are, and they last forever. Our physical bodies are only temporary representations of us, and they last for only a while. Their purpose is for fulfilling the reason our universe was created.

Making wars, mistreating one another, and doing criminal acts of all kinds are not a part of their purpose. Such behaviors mean we humans are living mainly our physical part of life and not much of our spiritual part of life which is the more important part of life.

14. **A Conclusion and a Look Ahead:** A conclusion is that the way we humans are living our lives is not compatible with how things work in the spirit world, and that is a major reason we are making a mess on Earth. This incompatibility developed because of the following:

> a. We have pretty much forgotten we have the spiritual part of life going on the same time our physical part of life is going on.
>
> b. By not being much aware and in touch with the spiritual

part of life we are unable to have good access to the wisdom that is formed and maintained there.

c. Thus, we humans overall unable to gain enough wisdom to figure out how to live our lives more constructively and more meaningfully. It should be easy to see how making wars, mistreating one another, and doing criminal acts of all kinds are not constructive or meaningful. But without enough wisdom, some of us seem to not even care that these acts are not constructive or meaningful.

According to the presentations in this book, we humans at one time in our distant past were well aware and in touch with the spiritual part of our lives. Evidence of this is indicated by how many species of animals and other creatures have retained their awareness with the spiritual part of their lives. It is likely that all living things were well aware of the spiritual part of life when they were first created, this includes us humans.

Thus, the challenge for us is to find a way to inspire us human to regain our awareness of the spiritual part of life and start behaving overall better and wiser. This is why References 1 and 2 and this book were written. As was said earlier, it might take hundreds and maybe thousands of years to achieve, if ever. However, we have to try and we need to start somewhere somehow.

2.3. The Need to Regain Our Awareness of The Spiritual Part of Life

As explained in Reference 1 and by the new spiritual model, our consciousness, intelligence, mental abilities, memories, senses, emotions, feelings, etc. all reside with our spirit in the spirit world and not with our brain in our physical world as commonly assumed.

We need to also understand that our ability to form wisdom also resides with our spirit and not with our brain, and wisdom is formed in the spiritual part of life, not in the physical part of life. If we are mainly aware of the physical part of our lives and we hope to find wisdom there, it is not going to work. We would be preoccupied with such things as competing with others in numerous ways, our status, our survival, our physical well being, our ego, etc.

We might be able to think about things and thus form wisdom while we are doing physical things. Therefore, we can argue that it is possible to find wisdom in our physical part of life. However, the fact that we have to think about things to form wisdom means we are forming wisdom in the spiritual part of our life. Thinking is done in the spiritual part of our lives, and thinking while doing something physical only proves that both parts of our lives are going on at the same time.

The following are only some of the many observations we can make or might have made that clearly indicate the spiritual part of life is real and exists:

1. The clearest indication that a person's mental attributes reside with the spirit is that the ghost of a deceased person retains all the mental abilities and memories of the formerly living person. A ghost is simply an image or a manifestation that is formed by the spirit of the deceased person. Thus, mental abilities and memories reside with the spirit and not with the brain. Otherwise the mental abilities and memories would be gone after the brain is dead.

I indirectly experienced manifestations of my mom's spirit four times after she passed away. Details are described in Chapter Two of Reference 1. The experiences are compelling because the manifestations did not come directly to me. In all four cases the manifestations came to a person not associated with my mom, and that person then conveyed the manifestations to me. This clearly indicates the manifestations were not figments of my imagination being fired up by my high emotional state at the time.

2. A very explicit example of how the spirit of a living thing has the "detailed construction plan" for constructing the living thing is how a chicken egg is able to be transformed into being a baby chick. This example was presented earlier in this chapter.

3. Rupert Sheldrake presented in his book "Dogs that Know When Their Owners Are Coming Home", Reference 8, numerous examples of how various animals and other creatures such as dogs, cats, horses, cattle, etc. are able to do extrasensory things such as **(1)** reading their owner's mind, **(2)** learning things telepathically from what other members of their species have learned, **(3)** finding their owners even when their owners were stationed in an unfamiliar place in another country, **(4)** knowing danger was about to happen and would try to stop their owners from going further, etc.

Something nonphysical exists that these animals are in touch with that enables them to do such things. Sheldrake suggests this something is a morphic field. The new spiritual model indicates this something is the spirit world and thus it is the spiritual part of life. Sheldrake's morphic field could very well be what I would say is the spirit world or the spiritual part of life.

4. Monarch butterflies returning from their annual migration are able to find the exact same grove of trees from which their earlier members took off to start their migration. The returning members are four generations removed from the members starting their migration. They went through four generations during their

journey. How the returning members are able to figure out which grove of trees the starting members started from could also be explained by the knowledge and wisdom that are a part of the spiritual part of the life of those butterflies.

5. Similarly, salmons when they are ready to spawn would instinctively know they need to swim upstream against water currents in exactly the right river to return to exactly the same place where they were hatched. This is after having spent years out at sea.

6. Newly hatched giant sea turtles would instinctively know they must head toward the sea immediately to help themselves survive.

According to the new spiritual model, instincts consist of spiritual signals the spirit world is issuing to living things that rely on the spirit world to do a good part of their thinking for them to help them survive. Thus, instincts originate from the spiritual part of life.

The stronger the instincts, the greater would be the living things' reliance on the spirit world to do their thinking for them. We humans do essentially all of our thinking ourselves and therefore we have very weak instincts.

Having strong instincts means the animals and creatures are well aware and in touch with the spiritual part of their lives. This could also explain how all the animals and creatures mentioned were able to do extrasensory things. They were doing them by going through their spiritual part of life.

Since all living things on Earth are created by the same creator, I suspect all living things including us humans were well in touch with the spiritual part of life when they were first created. As time passes the various living things evolved. Those that could do much of their own thinking would become increasingly less in touch with the spiritual part of life.

In the case of us humans, we essentially do all of our own thinking. This would tend to have us become increasingly less aware and less in touch with the spiritual part of our life, because we became too preoccupied with the physical part of life. We became so fascinated with our technical advancements and all the good and bad ways of applying them that we tend to not think about the spiritual part of our lives. Thus, we eventually lost much of our awareness of the spiritual part of life.

We eventually think we have only the physical part of life, and we would thus live mainly the physical part of life. Not being well in touch with the spiritual part of life prevented us humans from gaining access to the wisdom that is naturally formed and maintained in the spiritual part of life. Consequently, we would do destructive and meaningless things such as make wars, mistreat one another, and do criminal acts of all kinds.

To inspire and motivate us humans to stop doing such bad things we need to regain our awareness of the spiritual part of life. The new spiritual model could help us do that by providing a more complete and more accurate modeling of how things work in the spirit world and thus also how things work in the spiritual part of life.

2.4. Man's Expressions Will Differ from the Creator's Original Thoughts

I often wondered about some of the inconsistencies that are embedded into some of our current conventional spiritual models, and why no one seems to notice them or points them out. Upon having formulated the new spiritual model, I could see how the inconsistencies might have come about.

The creator's thoughts are most likely free of inconsistencies, but when humans translated them into spoken language expressions, that is when inconsistencies can develop.

It is well known that spoken languages are unable to translate thoughts totally completely or absolutely accurately. A given thought translated into expressions with several different languages would come up with several different meanings even if the differences are small.

However, the differences could be quite large if the translators happen to be very passionate about a thought. For example, one translator might want a strong expression while another translator might want a mild expression.

In the case of translating the creator's thoughts into expressions that could be used to formulate a religious model, I can see the translator might want to make the expressions very strong with the good intention of trying to make the religious model very effective.

This could explain why some of the features of the more popular religious models are generally quite extreme as exemplified by certain ones of the following features that are commonly found in religious models:

1. God is all knowing and all powerful.

2. God is loving, caring, and forgiving

3. God love us so much he lets his only son die for our sins.

4. A kingdom of God exists in heaven.

5. We need to worship God.

6. A person is likely to sin.

7. A person needs to pray to God and ask the person's sins to be forgiven.

8. If a person worships God and has asked the person's sins to be forgiven, that person would go to heaven after that person dies.

9. If a person doesn't worship God, that person will go to hell and burn forever after that person dies.

The inconsistencies I see in these attributes are as follows:

1. If God is all knowing and all powerful why would he be so insecure, egotistical, judgmental, vengeful, and cruel to require humans to worship him or else those who don't would end up in hell and burn forever?

2. If God is a loving, caring, and forgiving entity, why would he create humans only to cast those who didn't worship him into hell to burn forever simply because they didn't worship him? The whole idea of worship or end up in hell sounds like terrible mistreatment of humans who did not choose to be created but was created by him anyway who is now mistreating humans in this manner.

3. If God is the source of wisdom, then why is he so vengeful of those humans who did not worship him? Being a role model of vengefulness is not a wise thing to be doing.

4. By his being so vengeful would cause some people to be vengeful as well. And people would then think it is OK to be vengeful and maybe even be perceived as being strong and be

respected by being vengeful. This could contribute to people making a mess on Earth.

5. By God creating hell in the first place indicates he has in mind an extremely cruel punishment for those who did not worship him simply because they did not worship him. That makes God appear to be quite controlling and wants everyone to fear him as a way of controlling everyone. This could be where the idea of "a God-fearing person" came from. Controlling people by fear is not a positive thing. God being a role model for this could be why some people on Earth use fear and mistreatment to control others. This comes up in its worst forms during wars even as late as in years 2022 and 2023 when Russia invaded Ukraine.

Such inconsistencies don't make sense when we consider that the creator has to be extremely intelligent and highly experienced in creating living and nonliving things in order to be able to create things as complex as our universe and us humans and other living things.

Thus, a possible conclusion is that **(1)** the creator's thoughts were highly unlikely to contain such inconsistencies. **(2)** Such inconsistencies developed when the creator's thoughts are translated into spoken language expressions by human translators. **(3)** The translators wanted a religious model to be very strong so as to make it very effective. **(4)** Thus, the translated expressions were made very strong, and this is when the inconsistencies developed.

By contrast, the new spiritual model provides a more complete and more accurate description of how things work in the spirit world than the existing religious models could. The new spiritual model reveals the following features about the spirit world, and they are generally very different from what the religious models described about the spirit world.

1. The spirit world is the creator of everything including our universe and everything in it.

2. Oneness would naturally and automatically develop everywhere in the spirit world. This makes the spirit world naturally loving, caring, non-egotistical, forgiving, and nonjudgmental.

3. The thoughts formed by the spirit world would naturally embody the attributes mentioned in Item 2.

This supports the idea that the inconsistencies in the religious models must have come from man's translations of the creator's thoughts into spoken language expressions.

The following provides a comparison between some **features of the religious models,** shown as follows and as numbered in bold, and the *alternative interpretation by the new spiritual model,* indicated by the heading in italics and in bold.

1. God is all knowing and all powerful.

Alternative interpretation by the new spiritual model: The spirit world contains orders of magnitude more pieces of knowledge than the spirit of a human would contain. Thus, by comparison the spirit world could be perceived as being all knowing and all powerful by us humans.

2. God created our world and us.

Alternative interpretation by the new spiritual model: As explained by the new spiritual model and in References 1 and 2, the spirit world is the creator of everything including our universe and us. The spirit world is extremely large such that it encompasses all perceptions of God. Thus, any perception of God would be a portion of the spirit world, most likely only a small portion.

3. God is a loving, caring, and forgiving entity.

Alternative interpretation by the new spiritual model: Because of the oneness that pervades the spirit world, the spirit world is naturally and automatically a loving, caring, forgiving, and non-judgmental entity

4. God needs to be worshipped.

Alternative interpretation by the new spiritual model: Life consists of a spiritual part and a physical part. Humans need to be more aware of the spiritual part and to live it more consciously in order to gain access to the wisdom in the spirit world. Humans need to gain access to the wisdom in order to figure out how to live their lives more constructively and more meaningfully.

5. A person is likely to sin.

Alternative interpretation by the new spiritual model: Humans tend to focus too much on the physical part of life such they could eventually forget they have a spiritual part of life. Consequently, this could be why humans behave overall so poorly.

6. A person needs to ask God to forgive sins.

Alternative interpretation by the new spiritual model: For humans to fulfill their primary purpose for being here on Earth, they need to behave in a manner that is constructive and meaningful.

Humans are not likely to behave in this manner if they forgot they have a spiritual part of life. Humans need to regain their awareness of their spiritual part of life and thus regain their access to the wisdom in the spirit world so as to be able to figure out to live their lives more constructively and more meaningfully.

7. If a person worships God and has asked the person's sins be forgiven, that person would go to heaven after that person dies.

Alternative interpretation by the new spiritual model: By doing what is described by the non-bold letters in Item 6, the spirit of the person would be at peace thereafter while his or her spirit would be a part of the spirit world forever after the body of that person dies in our physical world.

8. If a person doesn't worship God, that person will go to hell and burn forever after that person dies.

Alternative interpretation by the new spiritual model: If the person does not do what is described in non-bold letters in Item 6, then the spirit of the person could have some regrets that could bother his or her forever while his or her spirit is in the spirit world forever.

However, no other spirit in the spirit world will judge that person's spirit. Only he or she would be doing any judging if he or she chooses to do so. Alternatively, he or she could look toward the future in the spirit world and do things that are more constructive and more meaningful.

The modeling of how things work in the spirit world achieved by the new spiritual model is thus shown to be more complete and more accurate than the modeling achieved by the various religious models.

However, this does not mean religious models should be discarded. They are correct in their spiritual guidance, but they are also fairly incomplete and inaccurate. In that sense they could constitute a good first step in a stepwise progression that would be developed to help us humans become more knowledgeable about how things work in the spirit world and thus how life works.

2.5. The New Spiritual Model's Features that Highlight:
A New Way of Thinking
New Unconventional Nonphysical Concepts
The Spirit World Is the Creator of Everything
Why Knowledge Is Among the Most Powerful Things Possible

Spirituality is a part of everything that exists. Each thing is either a spiritual entity that is a part of the spirit world or would have a spirit that is a part of the spirit world enabling the thing to exist in whatever is the place in which it exists. That place could be a universe or some other kind of place. For simplicity, let's call all such places universes.

Each universe will naturally and automatically be different and unique. Why this is the case is explain in detail in Reference 1, and is briefly explained as follows.

The spirit world is a living and growing thing. Just as anything that grows, it will periodically develop an imbalance. The spirit world would then create a new universe to help reestablish reasonable balance. Each new universe would naturally be different and unique because the spirit world will grow and be different each time an imbalance develops. Thus, the imbalance would naturally be different and unique each time such that the universe needed to reestablish reasonable balance would be different and unique each time.

Universes could be different in countless ways. They could be physical or nonphysical. They could be made of different sets of physical or nonphysical materials and energies. They could have different dimensionalities. Universes with the same dimensionality could have different kinds of dimensions making up their dimensionality. Some universes would have gravity and some would not have gravity. Etc.

What is said so far is based on logic and is basic enough and general enough to not be limited to any particular universe or to any particular thing in any universe. Logic is applicable for anything. Thus, to take science into consideration in the formulation of the new spiritual model, scientific and engineering logic was applied throughout the formulation process. We will see various examples of how this was done as the presentation continues.

Thus, as mentioned earlier, the starting point for the formulation process for the new spiritual model was the undisputable and logical statement that **"something, somewhere, somehow knows how to enable our universe and everything in it to exist"**. This statement represents a new way of thinking about spirituality. It is very basic and general and could be used with physical concepts or nonphysical concepts.

It was used with nonphysical concepts to formulate the new spiritual model since the model was to describe how things work in the spirit world, and the spirit world is a nonphysical entity. Using nonphysical concepts would be compatible with modeling how things work in the nonphysical spirit world, whereas using physical concepts would not be compatible.

Using physical concepts could still work, but the modeling process would be more complicated, less complete, and less accurate because of the incompatibility. This is case for the conventional religious models and for the models formulated by physicists such as the one presented in Reference 3.

Several new unconventional nonphysical concepts were introduced in a logical manner in the formulation process for the new spiritual model as described in References 1 and 2, and are briefly listed as follows:

1. **Knowledge Comes in Pieces.** This why in school and colleges a subject could be taught with a sequence of lessons. Each lesson covers a certain number of pieces of knowledge. This is how various skills and abilities could improve with practice. Each practice session is an experience that generates new pieces of knowledge. Each increase in the number of pieces of knowledge means an increase in understanding in how to perform the skill or ability.

2. **Each Piece of Knowledge Is Naturally Spiritual and Nonphysical.** Thus, as explained in Reference 1, all existing pieces of knowledge together would form the spirit world. Accordingly, at any point in time, the spirit world would know how to do everything that is possible to do at that point in time, and be able to create everything that is possible to create at that point in time. This is because the spirit world would possess all the pieces of knowledge that exist at that point in time. Thus, it has to be the case that **the spirit world is the creator of everything.**

Therefore, as it is often said, **knowledge is the most powerful thing possible.** The new spiritual model confirms this as being the case as stated in the preceding paragraph.

3. **New Pieces of Knowledge Are Generated by Experiences and Only by Experiences.** Experiences could be gone through by anything physical or nonphysical, living or nonliving, in any

universe or place including within the spirit world.

Living things could form their own experiences or have experiences happen to them. Nonliving things could only have experiences happen to them. Every experience would generate some new pieces of knowledge and some already existing pieces of knowledge.

4. Only One Copy of Every Existing Piece of Knowledge Is Kept. Duplicates do not increase the state of knowledge for the spirit world. They would only form a clutter in the spirit world. Therefore, duplicates would merge and only one copy will remain.

5. **The Formation of the Spirit World.** Every existing piece of knowledge has the natural spiritual energy to form a connection of the first kind with every other existing piece of knowledge. This then draws all existing pieces of knowledge together to form a spiritual entity. This spiritual entity would naturally be the spirit world.

Thus, at any given point in time the spirit world would consist of all the pieces of knowledge that exist at that point in time all connected together with connections of the first kind.

6. An Existing Piece of Knowledge Would Form a Connection of the First Kind with Every Other Piece of knowledge. Thus, at any point in time the spirit world would consist of every piece of knowledge that exists at that point in time all connected together with connections of the first kind.

7. Any Two or More Pieces of Knowledge All Connected Together with Connections of the First Kind Would Form a Spiritual Entity. Thus, a given piece of knowledge could be a part of many different spiritual entities ranging in size from just two pieces of knowledge to all the existing pieces of knowledge.

The largest possible spiritual entity at any point in time would be the spirit world. At any point in time the spirit world would consist of all the spiritual entities that could be formed with all the pieces of knowledge that exist at that point in time.

8. **Oneness Naturally Develops in the Spirit World.** A given piece of knowledge would be a part of a number of different spiritual entities. A given smaller spiritual entity would be a part of a number of larger spiritual entities. Thus, each spiritual entity would naturally be directly or indirectly a part of every other spiritual entity.

This is how oneness would naturally develop everywhere in the spirit world. Thus, the spirit world is where oneness is formed and where oneness resides.

9. **Each Piece of Knowledge Issues Its Own Unique Spiritual Signal,** and that is how one piece of knowledge or any one spiritual entity could be identified as being different from any other piece of knowledge or any other spiritual entity.

10. **Spiritual Senses Are Formed in the Spirit World.** Spiritual senses detect spiritual signals in the spirit world much like how our five human major senses detect physical signals in our physical world.

11. **Each Living Thing Would Have Its Own Set of Spiritual Senses** just as how it would have its own set of major senses. A living thing would use its spiritual senses to sense and find spiritual things in the spirit world much like it would use its major senses to sense and find things in our physical world.

12. **The Origin of Life Consists of Consciousness, Intelligence, Curiosity, and Wisdom.** The four nonphysical spiritual qualities: consciousness, intelligence, curiosity, and wisdom (referred to as the **4Qs** in Reference 2) are qualities of life in that any living thing would have at least one of these four qualities. They are spiritual things just as how pieces of knowledge are spiritual things.

This suggests that the four qualities coming together and interacting together would form the origin of life. The origin of life has to be a spiritual thing since logically nothing else could exist spiritually or physically unless something spiritual and alive knows how to enable them to exist.

Thus, life has to originate from something somewhere somehow. These four qualities provide a logically possible spiritual way life could be initiated.

Since logically the origin of life has to exist before anything else can exist, the origin of life must be the entity that initiated the existence of the spirit world.

The four qualities having curiosity as being a part of it would naturally be curious. Its curiosity would lead to its forming early experiences, and the early experiences would generate the early pieces of knowledge.

Such early pieces of knowledge would initiate the existence of the spirit world since at any point in time the spirit world would be made of all the pieces of knowledge that exist at that point in time. This would be a possible way the four qualities could initiate the existence of the spirit world.

All these new unconventional nonphysical concepts are basic

and general and are not associated with only any one specific planet in our universe or with only any one specific universe. Therefore, they would be applicable for any universe including our entire universe and not just for Earth or for any other planet in our universe.

These concepts are the bases for all the features of the new spiritual model. The features are presented in greater detail in References 1 and 2 and are also briefly described later in this chapter.

2.6. Unconventional Concepts and Features Used for: Formulating the New Spiritual Model. Applying Scientific & Engineering Logic Throughout The Formulation Process.

The formulation process for the new spiritual model presented in References 1 and 2 was not repeated here. However, its nonphysical concepts and features are summarized in this section of this chapter: Some of the ones mentioned earlier in this book are included here in greater detail. A few such "stepwise progression", "diversity", and various others are discussed further later as the presentation continues because of their importance.

Concepts and features that have the potential to be confirmed with experiments and observations are indicated with a double asterisk **.

1. **Something, Somewhere, Somehow Knows How to Enable Our Universes and Everything in It to Exist:** This was the opening statement that initiated the formulation of the new spiritual model. This was the first example of how scientific and engineering logic was applied throughout the formulation process.

2. **Knowledge Comes in Pieces:** Knowledge obviously comes in pieces, and we humans intuitively knew it all along. As mentioned

earlier, this is why for example in schools and colleges subjects could be taught in a series of lessons with each lesson building upon preceding lessons. Also, because knowledge comes in pieces, we could apply knowledge learned from one subject to help us learn the knowledge of another subject.

Thus, "knowledge comes in pieces" was another example of the scientific and engineering logic that was applied in the formulation process of the new spiritual model.

3. **Characteristics of Pieces of Knowledge:** By analogy a piece of knowledge is like a word. It is neither right nor wrong and neither good nor bad, and this analogy goes much further. Examples are the following.

Several pieces of knowledge connected together forms a spiritual entity. Spiritual entities could be many things such as a spirit of a living or nonliving thing, the solution to a problem, or simply a thought. By analogy, several words connected together forms a sentence. Sentences could be many things such a definition, the solution to a problem, or simply a thought. More such examples are presented in Reference 1.

A specific piece of knowledge could be a part of an unlimited number of different spiritual entities just as a specific word could be a part of an unlimited number of different sentences. Thus, only one copy of any existing piece of knowledge is needed in the spirit world just as only one copy of any word is needed in a dictionary.

According to the new spiritual model, every existing piece of knowledge forms a connection with every other existing piece of knowledge. Thus, the new spiritual model is able to explain why a piece of knowledge generated in one area of specialty could be useful in another area of specialty.

By analogy this is also true for words; i.e., words learned from one area of specialty can be used in another area of specialty.

4. New Pieces of Knowledge Are Generated by Experiences and Only by Experiences. The number of new pieces of knowledge that could be generated is limitless just as the number of new words that could be formed is limitless. This is because the number of different experiences that could be form is limitless. Therefore, new words are constantly periodically being formed also because new experiences are constantly periodically being formed.

5. Every Experience Is a New Experience Because No Two Experiences Could Be Exactly Identical, not even routine experiences we go through everyday such as brushing our teeth. Minor differences each time could include water temperature, how we do brushing strokes, the tooth brush being slightly more worn each time, the amount of tooth paste used could be slightly different each time, etc.

Thus, every experience would be new and would generate new pieces of knowledge. Some already existing pieces of knowledge would also be generated.

6. Only One Copy of Each Existing Piece of Knowledge Is Kept. Duplicates do not increase the state of knowledge in the spirit world. Thus, duplicates would merge and only one copy will remain.

7. Each Existing Piece of Knowledge Forms a Connection of the First Kind with Every Other Existing Piece of Knowledge: At any point in time every existing piece of knowledge has the natural spiritual energy to form a connection of the first kind with every other existing piece of knowledge so as to express every meaning that is possible to express with all the pieces of knowledge that exist at that point in time. We will see as the presentation continues that this spiritual energy every piece of knowledge possesses would naturally give the spirit world its creative power.

By analogy every word could be perceived as having an urge to connect with other words so as to express every possible meaning that could be expressed. This enables, for example, an author to create articles, stories, documents, expressions of feelings and emotions, etc.

8. Each Piece of Knowledge Issues Its Own Unique Spiritual Signal. This enables each spiritual entity to have its unique combination of spiritual signals and thus be easily identified and easily found when we are looking for a specific spiritual entity within the spirit world. This how, for example, we find thoughts. The process of finding thoughts goes very quickly, and that is how we are able to carry on conversations with others.

By analogy, each word has its own unique meaning. This enables each sentence to have its unique combination of meanings and unique expression such as a specific thought. Thus, in this respect, we could say a word in our physical world is a manifestation of a piece of knowledge or spiritual entity in the spirit world. In other words, a word exists in our physical world because a piece of knowledge or a spiritual entity in the spirit world enables the word to exist in our physical world.

9. A Thing Exists In Our Physical World Because It Has a Spirit In the Spirit World Enabling It to Exist. This is a very basic and fundamental concept regarding how life works in our physical world. This will become more apparent as the presentation continues. The new spiritual model is able to capture and describe this feature.

10. Life in Our Physical World Consists of a Spiritual Part and a Physical Part: The fact that we have a spiritual part of life and a physical part of life going on at the same time was briefly explained earlier in this chapter. This also was captured and described by the new spiritual model.

Finding thoughts, solutions, designs, concepts, etc. in the spirit world is a part of the spiritual part of life that each of us has going on in the spirit world at the same time our physical part of life is going in our physical world. Thus, this is a clear example of how we have a spiritual part of life going on in the spirit world the same time our physical part of life is going on in our physical world.

It is also an example of how the two parts of life are naturally interacting with one another and how such interactions are an important feature of how life works.

All this applies specifically to our particular physical universe and also in principle to other physical universes. In the case of a nonphysical universe, the concept of a spiritual part of life still applies. However, the other part of life would be nonphysical instead of physical. Thus, both the spiritual part of life and the other part of life would be nonphysical, but interactions between the two parts of life must still be a necessary feature of how life works in that case.

11. How the 4Qs would Form the Origin of Life: Explained in detail in Reference 2 is how the four qualities: consciousness, intelligence, curiosity, and wisdom are nonphysical spiritual qualities that when combined and interacting among each other would form the origin of life, according to the new spiritual model.

For short, these four qualities are referred to as the **"4Qs"** in Reference 2. The explanation presented in Reference 2 regarding how this works is updated by the more logical explanation presented earlier in this chapter of this book and in the discussion given as follows.

This is another one of the many new unconventional nonphysical concepts that is a part of the formulation process for the new spiritual model. While this is a concept, it is also a feature of the new spiritual model.

These are the four qualities that any living thing would have at least one of. For example, plants would have consciousness in the form of awareness of where sunlight is and where water is. Plants have some amount of intelligence by knowing how to grow and move to get sunlight and find water. Some even knows how and when to produce toxins after being attacked that could prevent predators from further attacking it. Humans and most animals and creatures have all four of these four qualities.

Thus, in a "back engineering" sort of way, it makes sense conceptually that the four qualities are spiritual qualities that when combined and interacting together would produce spiritual life and therefore would be the origin of life.

And as implied earlier, nothing spiritual or physical could be alive unless there is a living spiritual entity in the spirit world enabling it to be alive. Accordingly, to be the origin of life, the 4Qs would have to be able to be this entity, and based on the discussion, the 4Qs is able to be this entity.

As such, it must also be the entity that initiated the existence of the spirit world. How the origin of life could initiate the existence of the spirit world is as follows.

Basically, the curiosity quality of the four qualities would be curious enough to form experiences that all four qualities would go through to see what happens. As it turns out, any thing going through any experience would generate new pieces of knowledge. Thus, the early new pieces of knowledge would initiate the existence of the spirit world since at any point in time the spirit world is made of all the pieces of knowledge that exist at that point in time.

In fact, curiosity is still the spiritual quality that entices current intelligent living things to form new experiences that they could go through and see what happens. This is how intelligent living things do explorations and how we humans do research. This suggests that curiosity must have been the spiritual quality of the four spiritual qualities that got together with the other three spiritual qualities to form new experiences way before anything else exists besides them.

Perceiving the origin of life as consisting of these four spiritual qualities and having the spiritual powers as needed to be the origin of life is just as valid as perceiving the spirit world as consisting of pieces of knowledge and having its spiritual creative powers.

Pieces of knowledge naturally connecting and interacting together to form the spirit world is one of the new and unconventional concepts that are a part of the new spiritual model. The four qualities mentioned naturally connecting and interacting together to form the origin of life is another new and unconventional concept that is a part of the new spiritual model.

Additional other new and unconventional nonphysical concepts are formed and are applied throughout the formulation process of the new spiritual model.

12. **The Spirit World:** Because every existing piece of knowledge forms a connection of the first kind with every other existing piece of knowledge, all existing pieces of knowledge would be connected together and therefore all would be located in one place. That place would be called the spirit world.

Calling that place the spirit world is appropriate because as explained in Reference 1 all spirits of living and nonliving things are spiritual entities, and all spiritual entities are formed in the place where all existing pieces of knowledge reside. Thus, that place could be naturally called the spirit world.

At any point in time the spirit world would be made of all the pieces of knowledge that exist at that point in time. Therefore, any creative power the spirit world has have to come from the power each existing piece of knowledge has to form a connection of the first kind with every other existing piece of knowledge.

Therefore, at any point in time, all possible spiritual entities that could be formed would automatically be formed with all the pieces of knowledge that exist at that point in time. Each spiritual entity is the spiritual form of something, and some of them could enable things to exist in our physical world. This is how the spirit world is the creator of everything that exists in our physical world.

By similar reasoning, the spirit world would also be the creator of every other universe and of everything that exist in every other universe.

The spirit world is a growing and learning entity because new experiences are constantly being gone through by something somewhere somehow and thus new pieces of knowledge are constantly being generated and added to the spirit world. Thus, the spirit world is constantly growing and is also essentially constantly learning.

Anything that is growing and learning is a living thing. Accordingly, the spirit world is a living thing. This is how it is able to create living and nonliving things. It takes a living thing to create living things.

13. **The Spirit World Is Never Complete or Perfect.** Thus, nothing it creates could ever be complete or perfect: The spirit world could never be totally complete or absolutely perfect, because it could never possess every possible piece of knowledge that could be generated. This is because there is no limit to the number of new experiences that could be formed and thus there is no limit to the number of new pieces of knowledge that could be generated.

Because the spirit world is never complete or perfect, and it is the creator of everything, everything that exists would also never be complete or perfect. Our experiences indicate this is the case. However, everything being incomplete and imperfect is also a good thing.

14. **Living Things Are Alive Because They Are Incomplete and Imperfect:** The incompletion and the imperfection of living things enable living things to be living things. This is because by being incomplete and imperfect, living things have room to grow and to learn. Otherwise if something were complete and perfect, it would have nothing more to learn and nothing more to become. This means it could no longer learn or grow, which means it is no longer a living thing.

Nonliving things are also incomplete and imperfect since they are also created by the spirit world. This means nonliving things can in effect always learn and grow to be better. That is why we humans are constantly developing better and more advanced versions of just about anything we make. Examples of things that get better and more advanced with each subsequent model are cell phones, computers, automobiles, airplanes, washing machines, central heating and air conditioning systems, etc.

As explained in Reference 1, nonliving things have elemental consciousness and elemental intelligence. Therefore, they have elemental life. Elemental life is not the same as being alive. It is more like how an automatic switch knows when to activate, how chemicals know when to react, how ice knows when to melt and how water knows when to freeze. Etc.

Thus, for a living thing to be living things, its vital organs and various body parts need to be made of materials with elemental life. In other words, life is made up of a countless number of elemental consciousnesses and elemental lives all combined together and interacting together in a coordinated manner.

15. Religious Models are Correct But Incomplete: When various religious models depict God as being all powerful and all knowing and therefore perfect, then according to the earlier discussion, God would not be a living entity. This does not mean the various religious models are incorrect. It only means they are incomplete, and also inaccurate.

The religious models are correct regarding their spiritual guidance and are complete enough to inspire a lot of people to behave well. But, they are not complete enough or accurate enough to inspire all people to behave overall better and wiser. For example, we humans overall still make wars, mistreat one another, and do criminal acts of all kings.

This indicates we humans need something more complete and more accurate in order to be more effective in getting people to behave overall better and wider

The new spiritual model was formulated in a manner that enables it to provide a more complete and more accurate modeling of how things work in the spirit world and therefore how life works. A new way of thinking was applied along with new unconventional nonphysical concepts in its formulation process

By using nonphysical concepts instead of physical concepts as done to formulate existing spiritual models, the formulation process if much more compatible with the nonphysical nature of the spirit world. The formulation process also took science into consideration.

Thus, these two differences between the new spiritual model and the existing conventional spiritual models naturally make the new spiritual model more complete and more accurate.

16. The New Spiritual Model Was Never Meant to Replace Religious Models. Instead it could supplement the religious models. Therefore, the new spiritual model was able to accomplish the following compared with existing conventional spiritual models:

a. The new spiritual model is analytically and logically based instead of being faith based. Thus, it is likely to resonate better with today's more educated and more analytical population.

b. The new spiritual model is nonjudgmental which is how I think spiritual models should be. More specifically, they should be able to provide good spiritual guidance, and it should not be their place to be judgmental about it.

c. The new spiritual model used very basic and general non-physical concepts that would be applicable for our entire universe instead of being most likely applicable only with us humans on Earth.

The incompleteness of our religious models is also why we humans have so many different religious models and why some religious models have thousands of denominations. This wide range of choices is bound to be an attempt at compensating for the religious models' incompleteness.

17. A Stepwise Progression Could Help People Gain a More Complete and More Accurate Understanding about How Life Works. The first step could consist of the religious models. The second step could be the new spiritual model and the physicist's model. Subsequence steps would be models that are increasingly more complete and more accurate.

Stepwise progression has never been used to improve our spiritual models, and this could be why our existing conventional spiritual models have been so ineffective in getting mankind to behave better and wiser. Stepwise progression has proven to work great for our technical models and for the physical things we use in our physical lives.

Stepwise progression has proven to work well also for our educational system. Children would begin with preschool, followed by kindergarten, then first grade, second grade, etc. It is also how we bring up our children. It is also how practice improves our skills and abilities.

It is time we apply stepwise progression to continuously improve our spiritual models.

18. The Spirit World Is Likely to Have Been Initiated Multiple Times: Because the spirit world is made of pieces of knowledge, it has to have a beginning since pieces of knowledge do not exist without being generated first. Pieces of knowledge have to be generated by something going through experiences.

In Reference 2, the origin of life, the **4Qs**, was postulated to be the entity that could generate the first pieces of knowledge and therefore the 4Qs is the entity that could initiate the existence of the spirit world. How the 4Qs are able to generate the first pieces of knowledge was explained earlier in this section of this chapter.

It is possible the 4Qs was able to do a lot of things we humans couldn't even imagine. It is also possible that doing those things would keep the 4Qs a living spiritual entity, because those things would somehow be adding to the 4Qs and thus make it a growing and living entity.

By being a growing entity means the 4Qs must be an incomplete and imperfect entity, because at any point in time it could become more than it is at that point in time. This then means anything the 4Qs could bring into existence would also be incomplete and imperfect. Thus, being incomplete and imperfect is a natural attribute of being a living thing from the very beginning, just as it is with the origin of life, the 4Qs.

The various things the 4Qs would do would be instigated by the curiosity quality that is a part of the 4Qs. Thus, the curiosity quality would constantly entice the rest of the 4Qs to join in with the curiosity quality to explore various different things.

At one point in time, the curiosity quality must have gotten the 4Qs to see what happens when it forms an experience and go through it. The result was that doing that would generate pieces of knowledge. Then the consciousness, intelligence, and wisdom qualities would come to realize that pieces of knowledge could do a whole lot of new and interesting things. Thus, the 4Qs decided to initiate the existence of the spirit world to see what it could do.

This is strictly my speculation, but it is also logical and reasonable. Therefore, it could be real. If and when we humans learn how to do things by going through the spirit world, we will likely be able to see if this speculation is true or not. And if it is not true then we are bound to find out how the actual way the spirit world got initiated.

Among the new things the spirit world could do was that if the 4Qs could generate enough new pieces of knowledge and add them to the spirit world, the spirit world could become a living thing by being able to start forming new experiences and generating new pieces of knowledge on its own. In addition, it would be able to start creating living and nonliving things on its own.

Thus, the 4Qs realized "knowledge" could be a very interesting and powerful thing. The 4Qs then became curious about what can happen if it enables the spirit world to grow and be a living thing as explained, and this could be how the spirit world became what it is today.

Another thing the 4Qs found out was that anything that has a beginning and could grow to become self sufficient would have to go through a period in which it changes from being depended on its creator to being viable on its own. The new living thing is especially vulnerable during this transitional period and it could fail during the transition.

Thus, it is possible the spirit world was initiated multiple times before a version of it managed to survive this transitional period. Each version would be different because the pieces of knowledge that initiated it would be different. Some combinations of pieces of knowledge would know how to survive the transition better than others.

The version that survived this transitional period would thus become the spirit world that exists today. This version would thus also be able to create living things in which most could survive their transitional period. This could partly explain why so many species of living things exist on Earth. It is also most likely a great many species of living things would also exist elsewhere in our universe and in other universes.

19. **The Reason Our Universe Was Created and Why We Have a Primary Purpose for Being Here on Earth:** These two topics are explained in detail in Reference 1 and also briefly explained in several places in this book.

20. **Connections of the First Kind:** Each existing piece of knowledge has the spiritual energy to form a connection of the first kind with every other existing piece of knowledge. Any two or more pieces of knowledge connected together would form a spiritual entity. Thus, at any point in time, every spiritual entity that could be formed with all the pieces of knowledge that exists at that point in time would automatically be formed at that point in time.

Some of the spiritual entities would be capable of serving as the spirit for a living or nonliving thing in a universe. Thus, in this sense, this spiritual energy each piece of knowledge has is what gives the spirit world its power to be the creator of everything. This includes being the creator of every other universe besides ours.

It is often said that knowledge is the most powerful thing that exists, and according to the new spiritual model this is true.

21. **Spiritual Entities:** At any point in time, any two or more pieces of knowledge connected together with connections of the first kind would form a spiritual entity. The sizes would range from two pieces of knowledge to the largest possible being made up of all the pieces of knowledge that exist at that point in time. The largest possible spiritual entity would be the spirit world at that point in time.

Everything that exists in the spirit world is in the form of a spiritual entity. This includes such things as thoughts, solutions to problems, concepts, songs, complete stories, spiritual senses, etc. This also includes spirits of living and nonliving things ranging from tiny quantum particles to universes.

Some spiritual entities are thoughts. Since the structure of every spiritual entity is the same as the structure of thoughts, every spiritual entity is though-like. This means everything in the spirit world could be perceived as being various forms of thoughts.

The implications of this might be further explored in the future. My intuition tells me this could reveal things that are beyond what we have been able to imagine so far and that they are just as real as the things that has been revealed by the new spiritual model so far.

22. **The Spirit of a Living or Nonliving Thing Enables the Thing to Exist in a Universe by Essentially Thinking It to Exist:** Any thing that exists in a universe exists because it has a spirit in the spirit world enabling it to exist in that universe. Every spirit is a spiritual entity and every spiritual entity has the same structure as a spiritual entity that is a thought.

Therefore, in a sense, this means the spirit of a thing is enabling the thing to exist in a universe by essentially thinking it to exist in that universe.

This could explain why a scientist studying the smallest possible particle once said that when we get down to that small a level of existence, things seem to take on the qualities of a thought.

A related topic will be covered later in this book when quantum particles are discussed.

23. **Not every spirit in the spirit world could be translated into something that could exist or to be expressed in every universe:** Every universe is different and unique. Thus, for example, a spirit of a physical thing could not be translated into something to exist in a nonphysical universe. A spirit of a four-dimensional thing could not be translated into something to exist in a three-dimensional universe. Etc.

However, spirits of thoughts and spirits of entities made of thoughts could be translated into entities that might be expressed in any universe. Such entities could be, for example, certain stories, articles, documents, etc.

We might then ask: could the spirit of a story about events in a four-dimensional universe be translated into the story and the story can be expressed in a three-dimensional universe? The answer is yes. We humans have science fiction stories about life in universes that are different from our universe. It simply requires an author's imagination to write such a story and the imagination of the reader to follow it.

On the other hand, the spirit of a piece of music might not be translated into a musical entity that could be expressed in every universe. The musical entity might be expressed in a physical universe, but not sure if it could be expressed in a nonphysical universe since music is made of sound waves, and sound waves are transmitted through physical media. Perhaps the musical entity could be expressed in the form of a thought in a nonphysical universe.

24. **The Oneness that Pervades the Spirit World and the Attributes We Are to Address as Part of Our Primary Purpose:** Every piece of knowledge is shared among a large number of spiritual entities in the spirit world. Every smaller spiritual entity is shared among a large number of larger spiritual entities. The net result is every spiritual entity is directly or indirectly a part of every other spiritual entity. This is how oneness naturally and automatically develops through out the spirit world.

This means every human spirit has a major part of itself in common with every other human spirit. Therefore, our spirits would have the following attributes in our spiritual part of life:

a. We would not be judgmental of others.

b. We would not have opponents or win-lose competitions. We might have win-win competitions, and it would be for the benefit of everyone.

c. We would not be egotistic, controlling, possessive, or materialistic. Having an ego is only a concept and not a practice in the spirit world.

d. Fighting, killing, insulting, excluding, hating, etc. are only concepts and not practices.

e. Everything we do would be done together and it would be for the common good.

f. We would be living our lives constructively and meaningfully.

g. Anger, envy, hate, and fear would only be concepts and not practices.

h. The fear of dying would not exist because our spiritual part of life will last forever and be a part of the spirit world forever.

Thus, our primary purpose for being here on Earth is to do two things:

1. Help reestablish reasonable balance in the state of knowledge of the spirit world.

2. Help the spirit world learn how to carry out the physical part of our lives in a manner that is compatible with the following attributes.

Our spirit in the spirit world enables our physical body to exist in our physical world.

Therefore, our being a major part of one another in the spirit world is manifested in our physical world in numerous ways. For example, we share among one another: **(1)** more than 90% of our genes, **(2)** the same basic physical structure, **(3)** the same major senses, **(4)** the same kinds of feelings and emotions, etc.

Because we are separate entities in the physical part of our lives the physical attributes are different from the spiritual attributes of the spiritual part of our lives. The differences together with the new spiritual model enabled us to figure out our primary purpose for being here on Earth. This was covered earlier in this chapter.

Therefore, another way of looking at it is we are here to help the spirit world learn how to deal with the differences in a manner that is compatible with how things work in the spirit world.

This means our creator, the spirit world, intended us to be **"living our lives spiritually as being a part of one another and physically as being separate from one another."** And to do it in a manner that is compatible with how things work in the spirit world while at the same time helping to reestablish reasonable balance in the state of knowledge of the spirit world. To achieve this compatibility we need to carry out our lives in a manner that is constructive and meaningful.

25. Oneness Makes Friendships and Domestications Possible: The oneness among all living things is likely to be felt by other living things on Earth besides us humans. This could explain why some living things could be domesticated by humans. It amounts to such living things being able to form a friendly connection with humans and especially with a specific human individual such as the owner of the living thing.

This same kind of awareness of the oneness is likely to be how friendships are formed among living things besides among humans. This is also how former enemies could later become friends when the war is over and people have a chance to know each other on a personal basis.

Thus, friendships and domestications are possible confirmations that the oneness exists and that it exists in the spirit world since friendships and domestications have to originate in the spiritual part of our lives in order to be manifested in the physical part of our lives.

26. The Spirit of a Living or Nonliving Thing and the Spiritual Entity that Serves as Its Spirit: The spiritual entity that can serve as the spirit of a living or nonliving thing has to be large enough and complex enough to contain a smaller spiritual entity that is large enough and complex enough to be the spirit of a living or nonliving thing.

This has to be the case because of how reincarnation and evolution work. According to the new spiritual model, each time reincarnation occurs some amount of evolution will take place. This would bring into being a different spirit and thus a different living thing with each reincarnation.

A different portion of the larger spiritual entity would be employed to form the different spirit. To accommodate such differences each time reincarnation occurs, there has to be a spiritual entity that is involved that is larger than any of the possible spirits that could be formed to house the spirit that is formed with each reincarnation.

Thus, it is the larger spiritual entity that reincarnates such that it is the spirit that is formed within it that evolves and is more advanced than any of the preceding spirits that were formed within the larger spiritual entity.

27. The Spirit of a Nonliving Thing: A spirit of a nonliving thing is smaller and simpler than a spirit of a living thing. The spirit would not be housed within a larger spiritual entity because reincarnation and evolution would not happen to nonliving things. If a nonliving thing were changed, its spirit would change without going through reincarnation or evolution. Living things can form their own experiences and go through them. Nonliving things can also go through experience, but the experiences would be imposed onto them, usually by a living thing. An experience a nonliving goes through would likely result in some changes to the nonliving thing. The change would happen first with its spirit and then it would be manifested as a physical change on the nonliving thing.

28. **Reincarnation and Finite Lifespan Interactions:** As a living thing goes through experiences and generates pieces of knowledge, the pieces of knowledge would be added to the spirit and to the spiritual entity serving as the spirit.

Thus, both the spirit and the spiritual entity serving as the spirit would grow, evolve, and spiritually advance. But, the physical entity of the living thing residing in a universe would remain essentially the same and not physically evolve or advance. This and memories and habits contained in the spirit could get in the way of forming the experiences that could more effectively produce the next growths in the spirit and the spiritual entity serving as the spirit. Thus, the advancement of the entire species could be hindered in this manner.

To minimize such holding back of advancements for the species, living things are likely to be purposely designed to have a finite physical lifespan. By essentially "retiring" the old spirit to allow the now more advanced spiritual entity that was serving as the old spirit to now form a brand new spirit with no old memories and old habits to get in the way. Thus, the next round of advancements could now be more easily made.

This would also enable a new and more advanced and evolved physical entity to be formed.

By the new spiritual model describing this and also describing how we have both a spiritual part and a physical part of life going on at the same time, it becomes clear we humans can advance most effectively if we are living both the spiritual part and the physical part of our lives equally..

Therefore, the fact that we have been living mainly the physical part and not much the spiritual part could seriously hinder our overall advancement progress. This means the human species could have been much more advanced than we are now had we been living our spiritual part of life comparably with how we have been living our physical part of life.

This means we humans are way behind where we should be in terms of our overall advancement. The appearance of UFOs and UAPs on Earth could very well indicate this to be the case. More specifically, highly intelligent living things in other universes that are coexisting along with our universe have advanced to where they are able to do universe travel by not being hindered by behaving poorly. Thus, they are able to visit Earth in the form of UFOs and UAPs.

How such visitors would appear to us on Earth as UFOs and UAPs is explained in detail later in this book as the presentation continues.

We humans should have been able to do universe travel by now as well. But, we wasted too much of our lives and resources making wars, mistreating one another, and doing criminal acts of all kinds instead. We have thus hindered our advancement progress and are now way behind the highly intelligent living things residing in other universes.

It is time for us humans to behave better and wiser and try to catch up with where we should be in our advancement.

29. **Evolution:** The new spiritual model indicates three kinds of evolutionary changes could happen. The first is the familiar "the survival of the fittest" kind. Evolutionary changes will happen to match environmental changes.

This kind of evolution could also result in the formation of new species evolving from existing species.

The second kind is when the living thing generates new pieces of knowledge that is then added to its spirit. Thus, the spirit grows, becomes more knowledgeable, and more advanced.

This kind works by reincarnation interacting with the finite lifespan of living things. How this works was explained earlier in this subsection of this chapter.

The third kind is something referred to as being "Tinkering by Mother Nature", as described in Reference 1. This kind appears to be the spirit world simply exploring new concepts and new ideas.

The new spiritual model explains "Tinkering by Mother Nature" as something the "curiosity" quality in the 4Qs (the origin of life) would do. It is the "curiosity" quality in the 4Qs that is curious about what would happen if the spirit world were to try doing something new and different just to see what happens.

Thus, the new spiritual model is able to come up with a plausible logical explanation for the "Tinkering by Mother Nature" phenomenon.

A living thing could have all three kinds of evolution going on at the same time.

30. **Retracing ****: According to the new spiritual model, once evolution has taken place in a universe, an evolutionary path is established and can be retraced later in a future universe.

How this works is that all the new pieces of knowledge generated by establishing a new evolutionary path would be available for later use by the spirit world. Therefore, now that the spirit world knows how to do that particular evolution, it would do it again very quickly. This would be how retracing would work according to the new spiritual model.

The retracing process could go so quickly that no fossils would be left behind and thus retracing is often mistaken as being sudden creation.

The fact that so many species of living things on Earth, including the human species, have no fossils to indicate their evolutionary path suggests that they all emerged on Earth through retracing.

Thus, a lack of fossils indicates retracing is a real phenomenon.

31. **Instincts:** Instincts are spiritual guidance signals issued to the spirit of a living thing and are coming from the rest of the spiritual entity that is serving as the spirit. The guidance signals are specifically for the current spirit and are formed by the experiences of all the previous spirits the spiritual entity has served as.

Thus, we could say instincts are from lessons learned by all the previous spirits served by the spiritual entity that is now serving as the current spirit and are available to help the current spirit avoid repeating the same mistakes or accidents. While the current spirit would not remember the lives of the previous spirits, instincts could give the current spirit a vague idea of what some of the previous spirits must have gone through.

Instincts tend to be stronger for living things that do less of their thinking for themselves such that the spirit world would be doing more of the thinking for them. Thus, it makes sense their instincts would be strong since any part of the spiritual entity that is serving as the current spirit that is not a part of the current spirit would be issuing strong guidance signals to the current spirit.

Conversely, this explains why living things such as us humans that do most of their thinking themselves would have weaker instincts. It is because the spirit world is doing very little of the thinking for us humans.

32. **Intuition:** Intuitions are spiritual guidance signals coming from the rest of the spirit world that is not a part of the spiritual entity serving as the current spirit. Such spiritual guidance signals are available for all spirits of currently existing living things regardless of the universe the living things reside in.

Intuition is from lessons learned by every spirit that has been formed since the beginning of the spirit world. Thus, intuition is very general and is likely to touch upon just about anything about life.

While instinct signals are naturally limited in scope and are directed to a specific spirit such that it could easily tune into them, intuition signals are naturally extremely broad and are not directed to any specific spirit. Thus, a typical spirit is less able to tune into intuition signals compared with how well it is able to tune into instinct signals.

We might then ask: would a living thing such as a colony of bees that depends a lot on the spirit world to do the thinking for it be well tuned into intuition signals? The answer is likely to be no, and here is the reason why.

In order to be tuned into intuition in an effective manner, the living thing has to "sort through" all the categories of intuition signals and identify the ones that are relevant to itself at that point in time. It takes a certain amount of consciousness, intelligence, and thinking for oneself to do this.

A living thing that depends a lot on the spirit world to do the thinking for it is not likely to have the level of consciousness, intelligence, thinking for oneself to do this. Therefore, such a living thing is not likely to be well tuned into intuition signals.

In the case of us humans, some people are good at "sorting through" the categories of intuition signals and identifying the ones relevant to them. Those are the people who could best tune into intuition signals and to benefit most from them. Apparently, not every person is able to do the sorting and identifying necessary to be well tuned into intuition signals.

People in general can do the sorting and the identifying to some degree. So, it becomes a matter of how skillful and how accurately a person is able to do the sorting and identifying. A low level of skill could lead to a person making wrong guesses. Also, nothing is perfect so that even a person who is usually skillful could still make a wrong guess sometimes.

33. **The Role of the Spirit and the Role of the Brain of a Living Thing ****: Our body will do the things our spirit wants it to do. Our spirit is where decisions are made regarding what to do, and our brain is essentially the computer control center that enables our spirit and our body to communicate with each other.

Thus, the following interactions among our spirit, brain, and body are going on constantly when we are awake.

Our spirit issues spiritual signals to our brain. Our brain translates the spiritual signals into physical electrical signals and sends them to our body to instruct our body on how to move.

Our body issues feedback physical electrical signals to our brain. Our brain translates the physical electrical signals into spiritual signals and sends them to our spirit. This lets the spirit know how well the body is following its instructions.

This also enables the spirit to make modifications as necessary to compensate for any deviation between the instructions and the actual movements. For example, deviations could be caused by injury, tiredness, medications, constrictions, heavy loads, etc.

This back and forth exchange of signals between spirit and body by means of the brain goes very fast and this is how our body would move as our spirit wants it to move.

34. How the Spirit Pilots the Body **: The back and fourth exchange of signals between our spirit and our body by means of our brain as described earlier is how our spirit is essentially piloting our body. This is very much like how the pilot of an airplane pilots the airplane.

The pilot issues human signals to the computer control center of the airplane. The computer translate the human signals into electrical signals and sends them to the airplane to have the airplane do what the pilot wants it to do.

The airplane issues feedback electrical signals through sensors to the computer. The computer translates the feedback electrical signals into signals humans can understand and sends them to the pilot.

This back and forth exchange of signals by means of the computer goes on very fast and this is how the pilot pilots the airplane.

This is analogous to the exchange of signals between our spirit and our body by means of our brain. Thus, our spirit is essentially piloting our body, and our brain is essentially the computer control center for our body.

35. The Size of the Spirit vs. the Size of the Brain **: The size of a human spirit is very big because a human is highly conscious, highly intelligent, often has a lot of things to do and to think about, and the human body can do a lot of things and do them precisely.

The size of the human brain is also very big because the human body is capable of doing a lot of things and can do them precisely, and the human body is fairly large so there is quite a lot of body to pilot.

However, the size of the spirit and the size of the brain are not necessarily proportional with each other for living things in general.

For example, the spirit of a very smart bird can be big because the bird is intelligent, can solve puzzles such as figuring out how to open locks and doors, knows when to drop nuts onto a roadway so that cars can run over and crack them, can learn to speak words, and can use tools, etc.

But the brain of the bird is quite small compared with the size of its spirit because the bird's body can do only a very limited number of things and the body is small so there isn't much body to pilot.

Therefore, a very smart bird will have a very small brain compared with the size of its spirit.

On the other hand, a whale has a relatively large brain even though its body could do only a limit number of things. But, its body is very large so there is a lot of body to pilot. Therefore, the brain has to be quite large.

36. **Body Intelligence vs. Brain Intelligence ****: A living thing has body intelligence in addition to brain intelligence. In the case of a human body, all the vital organs could function in a manner that can keep the body alive even if the person is brain dead. Body intelligence is keeping the vital organs functioning properly.

This means even when the brain is not dead the spirit has a direct communication with the body and its vital organs without going through the brain in addition to having a communication path that goes through the brain. In other words, the body has some capability to understand spiritual signals or have some way to translate spiritual signals into electrical signals without the brain being involved in doing that.

This is logical since the body and the brain are made of the same kinds of basic body cells. This means all body cells have some limited capacity to translate spiritual signals into physical electrical signals and vice versa.

In cases in which a living thing such as a single living cell does not have a brain, the skin of the cell was determined to function as the brain for the cell. This suggests every cell making up a human body has its skin as its brain. This could be how body intelligence can exist for a human body.

The same would apply for other living things besides humans. For example, this might be how plants are able to have consciousness and intelligence such that they would know where sunlight is available and where water is available.

It would be interesting to see if we can design some method to detect such body cell consciousness and intelligence. Some ideas how this might be possible is explained later in this book when quantum particles are discussed.

Another example of body intelligence is when a person becomes very proficient at riding a bicycle, the bicycle seems able to stay upright without the brain being involved in keeping the bicycle upright. Body intelligence took over in that case.

37. **Connections of the Second Kind and How Auras Are Formed:** The connections of the second kind connect the spirit of a living or nonliving thing to its physical body or physical thing in our physical world. Each piece of knowledge making up the spirit forms a connection of the second kind going from the piece of knowledge to the physical body or physical thing. The connection carries the spiritual signal issued by the piece of knowledge to the physical body or physical thing to enable the body or thing to exist in our physical world.

As discussed earlier in this book, since the spirit is spiritual entity, it is structurally the same as spiritual entities that are thoughts. Therefore, the spirit is thought-like in its structure and in how it works. Accordingly, the spirit could be perceived as essentially "thinking" the body or thing to exist in our physical world.

The signals the connections of the second kind would be carrying are thus spiritual signals. Since the spirit of a person would include new pieces of knowledge generated by the experiences the person has gone through, the combination of all the spiritual signals carried by the collection of all the connections of the second kind is bound to embody information regarding the state of being of the person up to that point in time.

As the presentation continues we will see how it is that it is the spiritual signals going through the collection of all the connections of the second kind that produce the aura that could be seen surrounding the person's body, and that the aura would naturally contain information about the person and the history of what the person has been through up to that point in time and therefore the state of being of that person at that point in time.

As described in Chapter Two of Reference 1, I have a very special friend who could see and read auras. She was able to very accurately tell my state of being each time she sees me by looking at my aura. Another woman at a women's health fair was able to tell the state of being for my wife and me. She was roughly 90% accurate for both of us, and she was able to detect something about my wife that only my wife and I knew.

In my mind, this power of auras could potentially be put to good use in the medical care field as follows. An aura is three-dimensional since our body is three-dimensional. Therefore, and aura could provide a three-dimensional image of our entire body and its vital organs.

The amount of spiritual signals increasing with time as the body keeps going through experiences would not change the structural make up of the body or its configuration. But, the make up of the cells that make up the body would change. Sometimes the changes could make the body healthier and more resistant to heart attacks, strokes, and diseases, etc. and sometime the changes could do the opposite. It all depends on the experiences the person goes through whether they are healthy experiences or unhealthy experiences.

This implies that the state of being of the quantum particles that make up the cells that make up the body could contain information regarding the history of what the person has been through up to that point in time. That might open up a new way to diagnose and treat diseases. It would be non-invasive, could be more effective, and be easier to perform since we would be tackling the disease right at its core which is located at the interface between the spirit world and our physical world.

This is an example of how we are able to spiritually and technically advance at the same time by have a spiritual model that takes science into consideration and also a scientific model that takes spirituality into consideration.

38. **The Soul of a Living or Nonliving Thing ****: All the connections of the second kind collected together would form the soul of a living or nonliving thing residing in our physical world. Thus, the soul connects the living or nonliving thing with its spirit. The spiritual signals being issued from the spirit to the living or nonliving thing would go through the soul and enable the living or nonliving thing to exist in our physical universe.

The spiritual signals do that by enabling every quantum particle making up the living or nonliving thing to exist in our physical world. Thus, the soul would reside partly in the spirit world and partly in our physical world. Right at the location where the soul enters our physical world would be where quantum particles would be formed and located.

This could explain how a quantum particle would appear as an energy wave or at matter depending how it is observed. It would appear as an energy wave when observed from the spiritual side of the quantum particle that is facing the spirit world. It would appear as matter when observed from the physical world side of the quantum particle that is facing our physical world.

The soul would be connected with the quantum particle at the side of it that is facing the spirit world. That is why the quantum particle would appear as an energy wave. The energy we see would be that of the spiritual signal issued from the spirit of the living or nonliving thing.

The other side of the quantum particle would appear as matter because that side of it is in our physical world. The quantum particle would have to automatically change from being spiritual signal to being matter in order to be a part of our physical world. Thus, it is partly spiritual energy and partly matter.

Quantum particles are further discussed in a later chapter.

People who had out of body experiences said while they were floating around, they would see a silvery cord connecting them with their body. The part of a person that is floating around is likely the spirit of the person and the silvery cord is likely to be the soul of the person.

39. The Aura of a Living or Nonliving Thing Reveals the History of the Thing. **: As explained earlier, the spirit of a living or nonliving thing contains all the new and already existing pieces of knowledge that the living or nonliving thing generated up to that point in time. Thus, the spirit contains a complete history of what the thing has gone through up to that point in time. This means the spiritual signal issued by the spirit would also contain information regarding the history of the thing.

The aura of the thing is produced by the spiritual signal going through the soul of the thing. Therefore, people who could see and read an aura surrounding a person would be able to find out information regarding the history of that person.

Auras are further discussed in a later chapter.

40. The Spiritual Expression of a Living or Nonliving Thing: The spiritual expression of a living or nonliving thing is the combination of the spirit and the soul of the living or nonliving thing. At any point in time, the spiritual expression would always exist and be a part of the spirit world after the living or nonliving thing came into existence in any universe.

If the living or nonliving thing is existing in any universe then its soul would be residing partly in the spirit world and partly in that universe.

If the living or nonliving thing no longer alive and/or existing in a universe, then its soul would detach from that universe and will then exist completely in the spirit world and will stay attached to the spirit of the living or nonliving thing forever. The spiritual expression will remain as part of the spirit world forever thereafter.

This is how the spiritual part of life of a living thing would continue to go on in the spirit world forever after it's the part of life or existence in a universe is over.

If we are able to see things in the spirit world we would be able to see spiritual expressions of people and other living and nonliving things. If a person's physical life in our physical world is over, his or her consciousness would now be complete residing with his or her spiritual expression.

His or her spiritual expression would look like him or her at any age he or she wants, as explained in Reference 1. This is because the make up of his or her spirit would include how it was at any age he or she had been, and he or she could choose the version of his or her spirit to appear as in his or her spiritual expression.

41. Why and How Multiple Personalities Are Possible. Because a person's spirit grows with time, at any point in time a person's spirit would contain every version of itself that had existed up to that point in time. Therefore, theoretically the person could choose to have his or her consciousness reside with any one of those past versions.

Normally, a person would choose to be consciously at the latest version of his or her spirit so as to be aware of what is going on as life goes forward into the future, and thus he or she would be able to carry on with life accordingly.

However, it is possible for something extremely traumatic to happen at some point in time such that the person would choose to have his or her consciousness be temporarily with one of one or more past versions of his or her spirit for a while. But, in order to carry on with life, he or she has to have his or her consciousness be with the latest version of his or her spirit the rest of the time.

Such a past version of his or her spirit would have a personality that would be very different from his or her normal personality. This is because at that past moment in time he or she was trying to escape from his or her memory of a very traumatic event. He or she could do this by temporarily be someone else, and that someone else would have a very different personality.

Thus, as time goes by, now and then he or she might once again want to escape from his or her memory of the very traumatic event. Thus, now and then he or she would choose to have his or her consciousness reside with the past version of his or her spirit that would have him or her temporarily be that someone else.

I personally observed this happening with two individuals in my life, and the change in personality could be dramatic and be quite difficult for me to live with. This is mainly because I could never tell when the change would happen. In one case, I eventually understood why the multiple personality happened, and this helps to make coping with that person's second personality easier. After all, that person was only trying to survive.

To my huge surprise, an expert who could see and read auras was able to "see" and "read" the second personality of this person. The second personality was something only this person and I know about. The expert who saw it suggested that this person do journaling about it by describing it in writing and expressing that person's feelings about it in writing as a possible way to help resolve it or at least ease its effect on this person.

This was one of several instances that convinced me that auras are real and they really do contain a record of a person's history as they theoretically would, according to the new spiritual model. Thus, this was among the numerous evidences that confirmed the validity of the new spiritual model.

42. **The Power of Doing Journaling:** What is described in the immediately preceding topic confirms what I notice about how by my writing my books enables me to really tack down what the various concepts and ideas I have is about. If I don't write them down, somehow the concepts and ideas would remain fluffy and not be firm in their meanings and in what they are

about. Thus, writing my books is in effect my doing journaling regarding my thoughts, ideas, and concepts regarding spirituality.

Thus, in this case the new spiritual model was able to plausibly and logically explain why and how multiple personalities could happen.

43. We would be Able to Tell Which Universe a Thing Has Been in By Its Spiritual Expression. Each universe would have its unique conditions for the spiritual signals to satisfy in order to be able to enable a living or nonliving thing to exist in that universe. For example, the spiritual signals would have to be different to enable something to exist in a nonphysical universe compared with the spiritual signals that enable something to exist in a physical universe. The pieces of knowledge would be able to configure their spiritual signals such that such conditions are satisfied.

As indicated earlier, every universe is different and unique. Thus, the conditions the spiritual signals must satisfy would be different and unique for each universe. This means the nature of the spiritual signals being transmitted by the soul of a living or nonliving thing could reveal which universe the living or nonliving thing is residing in or had resided in.

After the part of life in a universe is over, the consciousness of the spiritual expression becomes complete tuned into the spirit world. The spiritual expression would then be able to interact with other spiritual expressions in the spirit world. It would be possible to tell which universe each of those living or nonliving things is residing in or had resided in by the nature of the spiritual signals its spiritual expression is issuing.

It would like how while we are alive in our physical world we can tell which country a person is from by the language he or she uses when he or she speaks.

44. The Ghost of a Living Thing, Nonliving Thing, or Deceased Living Thing **: Ghosts of deceased living things could be perceived in two ways. The soul having been partly in our physical world could partly return to our physical world to then appear as a ghost of the deceased living thing. The ghost would be in our physical world such that everyone has the possibility of seeing it.

The second way is some individuals and some creatures such as some cats are able to see things in the spirit world. Such individuals and creatures would be able to see spiritual expressions in the spirit world and the spiritual expressions would appear to such individuals and creatures as ghosts. Only individuals and living things that can see things in the spirit world would have the possibility of seeing the ghost.

Ghosts seen by the second way could be that of a deceased living thing or that of a still living thing since a spiritual expression could be that of a deceased living thing or that of a still living thing.

Theoretically, since spiritual expressions would exist for any living, nonliving thing, or deceased living thing, such ghosts could also be of living things, nonliving things, as well as of deceased living things.

I have a friend who has seen ghosts of a living person. I think the person whose ghost was seen must have been asleep at the time. Otherwise I would not think the spiritual expression of that person could be moving around independently from what the physical body of that person was doing.

During sleep, the spiritual expression could partly detach from the living body and move around in the physical world and be seen as a ghost in the physical world. The part that remains attached to the body would be the part that keeps the body alive.

This is also how out of body experiences could happen. Only part of the spiritual expression would detach from the body and have an out of body experience. The rest must remain attached to the body to keep it alive.

To determine if all that is said in this entire subsection of this chapter is true would require some research that could be quite interesting and revealing about how things work in the spirit world.

45. **Gradations Pervade the Spirit World ****: Every spiritual entity has a spiritual entity right next to it that differs from it by one piece of knowledge at a specific location of the spiritual entity. That second spiritual entity also has a third spiritual entity right next to it that differs from it by one piece of knowledge at the same location. This goes on and on at that same location and this would thus form a gradation running in one direction that corresponds with that one location.

The same thing could happen at every other location of the spiritual entity. Thus, the number of gradations that could form for a given spiritual entity would be equal to the number of locations, or the number of pieces of knowledge, making up that spiritual entity.

This means a given spiritual entity would be on as many gradations as the number of pieces of knowledge that it is made of. Since at any point in time the number of pieces making up the spirit world is countless, the number of gradation that is formed in the spirit world at that point in time would be countless, and the spirit world would naturally and automatically be filled with gradations everywhere and in every possible direction.

This also means every gradation could be a branch coming off of every other gradation. Therefore, for a given species of living thing, every variation that is possible for that species would have the potential of happening. In the physical part of life we can see how this is true for every species of living thing. In the case of creatures, this includes variations in size, color, color patterns, physical features, attractiveness, personality, preferences, awareness, intelligence, sexual inclination, etc.

46. Hardly Any Living or Nonliving Thing Would Be at an Extreme Location of Any Gradation **: For any given gradation, there could be essentially an infinite number of locations between the two extremes while there are only two extreme locations. This means hardly any living or nonliving thing could be located at an extreme location of any gradation.

Therefore, hardly any living or nonliving thing could be absolutely purely one way or the other. Thus, essentially every living or nonliving thing would be partly one way and partly another way. For example, for us humans on Earth, every person is partly straight and partly gay. No one is absolutely purely straight or absolutely purely gay. This can explain why gay behavior can happen in prisons among supposedly straight men.

A lack of sexual absoluteness also exists for various creatures besides humans. The point is that a lack of absoluteness with respect to any characteristic, feature, preference, etc. should be expected and accepted as simply natural for any species of living thing.

In particular for us humans, we need to understand that all varieties of variations that can happen for humans will happen. It is natural. Thus, individuals with variations should not be mistreated. After all, every one of us has some amount of numerous possible variations.

47. Gradations Are Natural and Important. Thus, Diversity Is Natural and Important. The important point for us humans to think about is that every variation in a given species is as important in the spirit world as every other variation. This is because every spiritual entity in the spirit world is as important as every other spiritual entity.

Logically, this is because every spiritual entity provides a different and equally important aspect of the spirit world and thus broadens the knowledge about how life works which then enhances the spirit world's ability to form wisdom

In particular for us humans, this similarly applies to every variation in the human species. Every race of humans and every variation of humans provides a different and equally important aspect of the human species and thus broadens the knowledge about how the human race functions. This then enhances mankind's ability to form wisdom.

It is particularly in this sense that diversity in all its forms in the human race could be perceived as a gift from our creator to mankind to give mankind the range of knowledge needed to form the wisdom that enables mankind to figure out how to be the best it could be.

Thus, every race and every variation of humans is important for mankind. This is why diversity in all its forms needs to be valued and embraced and be treated equally for the benefit of mankind.

We humans seem to understand the value of diversity in helping to preserve ecosystems. Thus, we work to preserve the diversity of other species of living things. We need to think similarly about the importance of diversity in our own species as well.

48. Therefore, Embracing and Valuing All Human Variations Is Definitely Smart and Wise Particularly at this Time**: The new spiritual model clearly indicates human variations are natural and zero variation is highly unlikely. In addition, human variations enable us humans as a whole to have a broader range of experiences and therefore also a broader range of knowledge and a deeper understanding about life. This enhances our ability to form wisdom regarding how we think and what we do.

Therefore, embracing and valuing all human variations would enhance our ability to achieve both spiritual advancements and technical advancements. This would be especially relevant for us humans at this time, because we humans have been way behind in our spiritual advancements compared with our technical advancements.

Our tendency to exclude individuals that are different from us is partly why we humans as a whole have been unable to form the wisdom necessary to figure out how to live life more constructively and more meaningfully. We tend to settle for less broadness in experiences and knowledge when we are making decisions, and thus we make decisions that are of lesser quality. This could add to our making a mess on Earth.

Real life examples showing the benefits of embracing and valuing diversity are presented later in the book.

49. Transplants **: It is a joint spiritual effort by the recipient's spirit and the donor's spirit to keep the body of the recipient alive. The donor spirit enables the transplanted organ to exist and functioning and the recipient spirit enables the rest of the recipient's body to exist and functioning, and both spirits interact together to keep both the donated organ and the recipient's body functioning in a coordinated manner such that the body can stay alive. .

Such interactions are possible because both spirits are more than 90% a part of each other and thus both are naturally able and willing to interact in this manner.

Thus, when we are doing any new medical procedure or doing any medical research on a possible procedure we are essentially finding a way to do it in the spirit world first. We would naturally do this by first imagining how it could be done and if it should logically work. The new spiritual model can now be a part of doing such explorations by providing a more explicit way to look at the logic of a potential procedure.

As time passes, new body cells are generated by the recipient body to replace any body cells that aged and died whether such body cells are part of the recipient body or part of the donated organ.

After enough time has passed, all the body cells of the donated organ would be replaced with body cells generated by the recipient body. The organ would then become completely made of body cells generated by the recipient body. At that time the donor spirit would no longer be involved with enabling the donated organ to exist and functioning. The task has been gradually turned over for the recipient spirit to perform.

Human spirits also share a lot of themselves with spirits of certain other living things such as various species of animals. This is why certain organs from certain animals are able to function when transplanted into a human body. Such transplants so far are usually done only for research explorations.

50. **The New Spiritual Model Could Easily Become a Natural Part of a Person's Everyday Life **:** This is because the new spiritual model could provide plausible spiritual explanation for just about any common everyday experiences and observations we can think of. Examples are: why we need sleep, why dreams tend to be surrealistic, how we find our thoughts, how we find solutions, how we have instincts and intuitions, why some people can see and read auras, how telepathic communications work, etc.

The spiritual model is nonjudgmental, does not impose rules and laws onto people, does not rely on faith for acceptance but is instead logical and analytical such that it is more in sync with today's more educated and more analytically oriented population.

Thus, the new spiritual model could easily feel comfortably and naturally a part of everyday life such that is would be easy to stay turned to it and its spiritual guidance every waking moment of everyday.

51. **Spiritual Senses Function in the Spirit World:** Spiritual senses detect spiritual signals in the spirit world much like how our five human major senses detect physical signals in our physical world.

The number of spiritual senses that could be formed in the spirit world is unlimited. New spiritual senses are constantly being formed as the spirit world continues to grow and new spiritual things are created that need new senses to be detected.

52. **Five Specific Spiritual Senses Give Us Our Five Major Senses ****: Five specific spiritual senses were extended into our physical world to give us humans our five major senses. Other spiritual senses were similarly extended into our physical world to give other living things their major senses, and their major sense could be different from those of humans. For example, sharks can sense magnet fields underwater, and platypuses can sense electrical charges underwater.

Because major senses are formed from spiritual senses, we humans and other creatures should be able to sense some things in the spirit world the way we sense things in our physical world. This could explain how some people are able to see and read auras. This could also explain how some creatures such as cats seem able to sense things we cannot sense.

This could also explain how some people can do remote viewing, see ghosts, see things while having an out of body experience, and have various other extrasensory abilities,.

53. **Telepathic Communication (TC): Various Forms of TC and Their Roles in the Formation of Wisdom ****: Telepathic communication is simply one form of the many natural ways communication takes place within the spirit world. It is how we get our instincts and our intuition and how remote viewing works. It is how cross learning can occur from one member of a species to another member of the same species. This is how we could sometimes find solutions to problems while asleep. Such various extrasensory abilities are also various forms of telepathic communications that takes place in the spirit world.

Every existing piece of knowledge has a connection of the first kind with every other existing piece of knowledge. Therefore, any signal issued by an existing piece of knowledge could find its way to another existing piece of knowledge through such connections. Basically, this is how all forms of telepathic communications work, and they all take place within the spirit world.

This is how wisdom could be formed in the spirit world. It is when every existing piece of knowledge contributes what it knows to the process of forming wisdom, and this is why the state of knowledge in the spirit world has to be kept reasonably balanced. Wisdom could be formed regarding any issue only when the knowledge on all sides of the issue is reasonably balanced.

This is also why every spiritual entity in the spirit world is as important as another spiritual entity. This then gets us back to why diversity in all its forms is an important feature in the spirit world, and we humans should value and embrace physical diversity in all its form in our physical world.

You might wonder why I emphasize the need to value and embrace diversity in all its form so much. It is because I conclude it is one of the most important things we can do to help us form the wisdom we need to help us stop making wars, mistreating one another, and doing criminal acts of all kinds.

I personally experienced receiving a message through telepathic communication on two occasions as follows:

> 1. At around age ten I was working in the garden behind the house late in the evening when it was getting dark. My mom was working in the front garden of the house at the same time. I heard her call me. So, I went to her and asked "Did you call me?" She sort of hesitated and then she said "No, but in my mind I was wishing you would come to me." She had called for me telepathically. That was a sweet moment to recall even now more than seven decades later.

2. Very recently at age 88 I was really stress out for a day and a half over something to do with my son. Then suddenly the stress was gone, and I knew at that moment what I was stress out about has just been resolved. Sure enough, a moment later I received an email from my son indicating everything is OK. So, it seems I have this telepathic connection with my children just as my mom and I had a telepathic connection between us.

54. Premonition and the Ability to Know the Future **: According to the new spiritual model, the spirit world would naturally and constantly make projections into possible futures based on past and present knowledge so as to be prepared for anything that has the possibility of happening. Some individuals and creatures are able to sense such projections and are thus able to know what will happen in the future, usually in the near future.

For example, some animals and birds are known to move to safer places days before a natural disaster is about to happen. Some animals could sense a bad situation is about to happen either soon, further down the road, or around the corner. The animals would do what they could to prevent their owners from being caught in the bad situation by doing such things as getting their owners to move out of the way or to go no further down the road. Several actual cases are described by Rupert Sheldrake in Reference 8.

55. Long-Range Premonitions That Are Written In the Bible **: The spirit world possesses all existing pieces of knowledge and therefore knows all the experiences that generated them. Thus, the spirit world could form some premonitions that are far into the future, perhaps hundreds of years into the future.

As mentioned earlier, the spirit world is constantly making projections into possible futures as part of its task to keep its state of knowledge in reasonable balance. Some projections could go far into the future.

For example if a civilization is being governed in a manner that is incompatible with how life works, it is easy to predict it will eventually fail, maybe even the particular year that it would fail. How long a civilization would last would depend on how compatible the way it is governed it with how life works. The greater the compatibility, the longer it will last.

As the presentation continues we will talk about the 87 ancient civilizations that were recently studied by Luke Kemp, Reference 9 that all eventually failed. In my mind, they most likely failed because the way they were governed was incompatible with how life works.

Some people, particularly some in our distant past, are likely quite aware of the spiritual part of their lives and therefore could sense some such long-range premonitions. This could be how some such premonitions got written into the Bible.

The fact that such accurate premonitions are in the Bible could support the possibility that some people in the past are aware enough of their spiritual part of life that they are able to sense and/or see things in the spirit world that a typical person would not be able to sense or see.

56. **Sleep Is a Restoration Process****: How It Works is as Follows: Our physical body exists in our physical world because our spirit in the spirit world is issuing spiritual signals into our physical world to enable our physical body to exist. These signals keep our body in the state of being that our body is supposed to be in. Thus, we are healthy mainly because these signals are keeping us in a healthy state of being.

While we are awake, all the physical signals that goes on in our physical world tend to drown out the spiritual signals that are keeping us healthy. Consequently, our body gets wear and tear. Thus, our spirit needs our body to sleep for a certain number of hours after a certain number of hours of being awake. This is to allow the spirit to restore the body back to the state of being it is supposed to be in. This is how our body recovers from its wear and tear and why we feel much better after a good night's sleep.

We need to sleep in a place where the physical signals from our physical world are absent or at least very low such that they would not interfere with the spiritual signals. Thus, the place is usually dark and quiet since light and noise are physical signals. The spiritual signals could then perform their restoration action.

The restoration is done in cycles probably because we humans have a habit of not sleeping long enough to enable the restoration to be complete. Therefore, the first cycle would restore the most important parts of our body such as our brain, heart, and lungs. Subsequent cycles would restore other parts of our body in accordance with their importance in keeping the body alive. That way we would still be able to function reasonably well during the day even if we are not completely restored.

It is commonly assumed the front cortex of our brain goes dormant when we sleep. But, according to the new spiritual model that part of our brain only appears to go dormant because our spirit is not issuing any signals to it to go into action.

If something important happens while we are asleep, we are able to wake up and respond right away because our spirit has issued a signal to the front cortex to get our body to respond. This means the front cortex remains ready to respond instead of going dormant when we are asleep.

57. Dreams Tend to Be Surrealistic for a Reason **: We are how we are because of all the experiences the spiritual entity that is serving as our spirit has gone through starting from the moment that spiritual entity has become able to serve as the spirit for living things. As the presentation continues, and as mentioned earlier, we will see how it is that the process of helping to restore reasonable balance for the spirit world is likely to involve a sequence of universes instead of just one universe.

This could explain why we humans are more advanced than any other living thing on Earth. It is because we humans emerged early in the sequence of universes and thus had a longer time span to evolve and advance than did any of the other living things on Earth.

This means the spiritual entity that is serving as our spirit has been through a lot of experiences that took place in a lot of past universes. As mention earlier, every universe is different and unique. Therefore, the experiences in one universe are not going to be like the experiences in any other universe. As a result the spiritual signals that enable us to exist in our physical world are made up of a combination of different signals issued by pieces of knowledge generated in a number of different universes.

Therefore, during sleep, the signals doing the restoration would be issued from our spirit to our body and the signals would induce dreams make up of a mixture of images from a lot of past universes. The individual images making up our dreams would all be difference from one another and be different from what we are use to seeing on Earth. Therefore, the composite image in our dreams would tend to be sort of Earth-like and would thus be surrealistic.

The surrealistic quality of the image does not bother us because our spirit knows why they are surrealistic. Our spirit is also able to let us know that when we are dreaming about something such as our house we would know it is supposed to be our house even though it does not look very much like our house because the image is surrealistic. We are not bothered by this because our spirit knows the image will be surrealistic.

58. Aging Still Occurs in Spite of Sleep Being a Restoration Process **: The wear and tear our body gets while we are awake during the day are caused by the experiences we go through during the day. These experiences generate new pieces of knowledge that get added to our spirit and would thus change our spirit. On one hand the change is growth and that is good. But, on the other hand the change also has an aging effect.

Thus, we can see how we are having physical wear and tear, spiritual learning, and spiritual aging all going on at the same time. The spiritual aging will be manifested as physical aging on the body.

During sleep the restoration process will restore mainly the physical wear and tear but could not do anything about the spiritual changes in the spirit. In other words, restoration would not and could not remove any new pieces of knowledge that got generated and added to our spirit. Thus, a bit of aging would occur with each day of life.

We seem to sense this intuitively as we tend to see older people as being more knowledgeable people and hopefully also wiser people.

59. Since Spiritual Expressions of Individuals Exist in the Spirit World Individually, Does Ego Exist in the Spirit World? The answer is, since everything is directly or indirectly a part of everything else in the spirit world, ego does not exist in the spirit world.

Other spiritual expressions would not judge us or think they are better or worse than we since every spiritual expression is a part of every other spiritual expression. All human spiritual expressions are more than 90% a part of each other.

In addition, a spiritual expression is simply a part of the spirit world much like how an arm or a leg is a part of our body. Thus, for example, our right arm would not judge our left arm or would think it is better or worse than our right arm.

We might ask: what if a person was a criminal while alive on Earth. Wouldn't other spiritual expressions think less of that person's spiritual expression after that person dies on Earth and has now become his or her spiritual expression? The answer is no because all human spiritual expressions are a major part of one another. Thus, thinking less of any spiritual expression would also be thinking less of him or her self. This is not going to happen.

What could happen is that the spiritual expression of a criminal might regret even being a criminal during its physical part of life, and this regret could stay with his or her spiritual expression forever. Alternatively, the regret could last for a while and then the spiritual expression would go on being a part of the spirit world and would participate with the rest of the spirit world in doing what the spirit world would be doing to maintain it own viability.

In other words, there is something more constructive and more meaningful to do than to continue regretting ever being a criminal since no other spiritual expression is going to judge what anyone has done while being a human on Earth.

Chapter Three

The Reason Our Universe Was Created and Our Primary Purpose for Being Here on Earth

In general, universes are created by the spirit world to help reestablish balances in the spirit world, and it is to be done in a way that is compatible with how things work in the spirit world.

Thus, our particular universe was created for a reason. And, it also follows that we humans have a primary purpose for being here on Earth, and it is to help our universe fulfill its reason for being created. The new spiritual model helped determine the reason our universe was created and therefore also our primary purpose for being here on Earth.

The new spiritual model explains that the spirit world is a living and growing entity. And, anything that grows would periodically develop an imbalance. In the case of the spirit world, any imbalance would has to be restored to reasonable balance, because the spirit world's state of knowledge has to be kept reasonably balanced in order to form the wisdom it needs to figure out how to maintain its own viability.

If not restored to reasonable balance fairly quickly, an imbalance would continue to worsen to where it becomes unstable and could not be restored. The state of knowledge would then become nonsensical. Then the spirit world would become nonviable and would vanish. And, everything it created would also vanish.

In a way, the life of the spirit world is quite restrictive, because it has only a spiritual life that is also only nonphysical. As explained earlier, oneness would automatically form throughout the spirit world. Accordingly, on its own, the spirit world could only form experiences that are spiritual, nonphysical, and in which things are directly or indirectly a part of one another.

In our lives we can see how anything that is a living thing will grow, and its growth would naturally periodically develop issues, and issues are essentially imbalances. Examples would include financial issues, children growing up and want more freedom, things wearing out and needing to be replaced, etc.

Such issues need to be "rebalanced" in order to continue to grow in a viable manner. This is particular the case when growth takes place in a restrictive manner. Financial issues are examples if the income is just enough to cover all expenses. Any issue has to be "rebalanced" quickly.

All kinds of imbalances could develop in the spirit world. Two kinds are most likely to be repeated periodically as follows:

1. Because the spirit world is nonphysical, on its own the spirit world would generate too many new pieces of knowledge to do with things being nonphysical.

2. Because oneness pervades the spirit world, on its own the spirit world would generate too many new pieces of knowledge to do with things being directly or indirectly a part of every other thing.

Whenever an imbalance developed that the spirit world could not resolve on its own the spirit world would design and create a new universe to help reestablish reasonable balance. Because of the two listed repeating natural causes of imbalances in the spirit world, a lot of the new universes are likely to be physical and/or such that things in the universes are separate from each other in addition to having various other features depending on other imbalances that had developed in the spirit world.

Our universe is one that is both physical and where things are separate from each other. So, right away we know our universe was created to include generating new pieces of knowledge that are to do with physical things that are separate from each other. Thus, our primary purpose for being on Earth includes helping to reestablish reasonable balance in these two areas and to do it in a manner that is constructive and meaningful. .

Since all humans have spirits consist of small portions of the spirit world, our physical part of life could be perceived as being here on Earth to help the spirit world learn how to live life physically as entities that are separate from each other while as the same time spiritually as entities that are a part of one another.

This means we are to do things as separate individual while at the same time treat each other with respect, care, empathy, compassion, etc. just like how we would be if we were actually a part of each other because spiritually we really are a part of each other. This will assure that the way we generate new pieces of knowledge will be compatible with how things work in the spirit world.

Perhaps the easiest way to assure how we generate new pieces of knowledge would be compatible with how things work in the spirit world would be to do it in a manner that is constructive and meaningful. Otherwise, as explained earlier, the new pieces of knowledge could not be used to reestablish reasonable balance in the spirit world. They would simply be added to the spirit world in a manner that could cause addition imbalances to develop, and additional new universes might be needed to be created to rebalance the imbalances.

Unfortunately, so far a lot of the new pieces of knowledge we human have been generating on Earth have been generated in a destructive and meaningless manner by making wars, or mistreating one another, or doing criminal acts of all kinds. Thus, overall, we humans have not been helping very much to reestablish reasonable balance in the spirit world.

Chapter Four

The Ingeniousness of the Creator of Everything as Pointed Out by the New Spiritual Model

4.1. The Creator's Ingeniousness is How Creating a Few Simple Basic Particles Can Form a Universe and All Living and Nonliving Things in It

The creator of everything was absolutely ingenious how it created only a few simple basic physical things such as electrons, protons, neutron, dark energy, and dark matter that could make up every living physical thing, every nonliving physical thing, every form of physical energy, and every celestial phenomenon that are a part of our physical universe. These same few basic things could also sustain all kinds of physical life in our universe.

Such creativity could only come from a lot of experience creating universes, all kinds of phenomena, all kinds of living and nonliving things, and all kinds of ways to sustain life. It also indicates the creator has amazing consciousness, great awareness of how things are interrelated, enormous ability to figure out things and resolve problems, and tremendous wisdom. These are attributes that enable the creator to be so ingenious.

All such attributes could largely come from the oneness that automatically forms throughout the spirit world such that everything is directly or indirectly a part of everything else. We could imagine how the overall interconnection among all things in the spirit world could enhance: overall consciousness, awareness of every thing and every phenomenon, the understanding of how all things are interrelated, and thus the formation of wisdom.

Conversely, the creator's amazing ingeniousness would therefore indicate oneness exists and pervades the spirit world since the oneness could promote the development of the mentioned attributes that give the spirit world its impressive ingeniousness. This in turn helps to confirm the validity of the new spiritual model since the model is able to describe the interconnectivity and thus also the oneness that automatically forms within the spirit world.

The new spiritual model is able to describe such attributes much better than religious models could. This indicates the new spiritual model is modeling how things work in the spirit world more completely and more accurately than the religious models were able to.

Physicists and various other scientists were also ingenious in coming up with physical concepts such as electrons, protons, neutrons, dark energy, dark matter, etc. that are able to define quite well what the simple basic things the creator created are and what they are able to do. This gave each simple basic thing its identity such that technical models could be formulated to describe all the physical things and physical phenomena that are possible to be produced with those simple basic things.

Examples include: all the atomic elements, all kinds of molecules, all kinds of molecular reactions, all forms of energies, all species of living things, all kinds of nonliving things, etc.

The creator also created all other universes. Some have come and gone, some still exist and are coexisting with ours, and more will be created in the future. The other universes are likely to be also made up of only a few simple basic things. However, each universe would be different and unique. Therefore, each would be made of a different set of simple basic things.

So far in this chapter we have been talking about the physical part of life. There is also the spiritual part of life. This is the part that the new spiritual model and the conventional religious models were formulated to describe.

It is interesting that in a parallel sense, the new spiritual model was able to describe how things work in the spirit world with only three simple basic unconventional nonphysical concepts as well. They are **(1)** pieces of knowledge; (i.e., knowledge comes in pieces), **(2)** connections of the first kind, and **(3)** connections of the second kind.

With just these three basic nonphysical concepts to formulate the new spiritual model, the model was also able to plausibly reveal for example: **(1)** what constitutes the spirit and the soul of a living or nonliving thing, **(2)** the reason our universe was created, **(3)** our primary purpose for being here on Earth, **(4)** how the spirit world can enable universes and things to exist by essentially thinking them to exist, **(5)** how our creator intended us to carry out our lives, **(6)** the reason we make a mess on Earth is because the way we are carrying out our lives is incompatible with how life works, and **(7)** numerous other attributes that are mentioned in this book, etc.

Going further, add on the fourth simple basic unconventional nonphysical concept, the 4Qs, and the new spiritual model was able to plausibly reveal: **(8)** a possible origin of life and **(9)** how the existence of the spirit world was likely initiated. The 4Qs was introduced and explained earlier in this book.

4.2. How Things Work in the Spirit World Is How Life Works for All Universes

Universes exist because the spirit world enables them to exist. Therefore, how things work in the spirit world would be the basis for how things work in the various universes, including how life works in the various universes. Since the new spiritual model is able to describe how things work in the spirit world, it could thus also describe how life works in the various universes.

For example, a living or nonliving thing in a universe would have a spirit residing in the spirit world enabling it to exist in that universe. Thus, the living or nonliving thing would have two parts to its existence; i.e., a spiritual part going on in the spirit world and a part that is going on in the universe in which it resides.

Since the spiritual part resides in the spirit world, how it works would be how the spirit world works. Therefore, how it works would be the same for all living and nonliving things regardless of the universe in which they reside. The part that takes place in their universe would be different for each universe since each universe would be different and unique.

For example, no matter the universe, the spirit of a living thing would still be piloting its living entity that exists in that universe. Thus, a living entity in any universe would need something like a brain that serves as the communications link between the spirit of the living entity and the living entity

Therefore, the new spiritual model could be referred to as being a model that describes how things work in the spirit world and also as being a model that describes how life works.

4.3. The Need for a New Way of Thinking And to Think Abstractly In Terms of Nonphysical Concepts

The more we are discovering about Earth and our universe, the more fascinating we find them to be. We are able to formulate technical models to describe our discoveries, and this is how we achieved our amazing technical advancements.

We are finding fascinating good ways to apply our technical advancements. Unfortunately, we are also finding bad ways to apply our technical advancements as well, and thus we are contributing to the mess we have made on Earth.

Our technical models are essentially documents of our technical advancements. Similarly, our spiritual models are essentially documents of our spiritual advancements. We have an enormous number of technical models, and their number keeps growing. But, we have relatively few spiritual models, and their number has not increased much for thousands of years.

Ideally, our spiritual advancements should keep pace with our technical advancements such that we would have the necessary spiritual guidance on how and how not to apply each increase in our technical advancements. But, our spiritual advancements lag far behind our technical advancements. Thus, without the necessary spiritual advancements to guide us, we applied our technical advancements in bad ways as much as in good ways.

This lack of wisdom is indicative of a lack of balance in our advancements, which means a lack of balance in our knowledge about how life works. The new spiritual model was formulated to help us find ways to gain the knowledge necessary to reach a reasonable balance between our technical advancements and our spiritual advancements.

As it turns out, this requires a new way of thinking and a need to think abstractly in terms of nonphysical concepts instead of thinking conventionally in terms of physical concepts as we have always done up to now.

Physical concepts work well for pursuing technical advancements because our technical advancements pertain mainly to physical things in our physical world. But physical concepts do not work so well for pursuing spiritual advancements because spiritual advancements pertain to how things work in the spirit world, which is a nonphysical world.

The new spiritual model is able to get us into a new way of thinking such as thinking abstractly in terms of applying nonphysical concepts. This is what we need to more effectively pursue spiritual advancements. This is because nonphysical concepts are compatible with the nonphysical phenomena that take place in the spirit world.

Our technical advancements and our good and bad applications of them are so fascinating that we tend to focus mainly on the physical part of life, and thus we eventually forgot we also have a spiritual part of life. The new spiritual model could reawaken our awareness of the spiritual part of life by helping us grasp a new way of thinking and helping us strengthen our ability to think abstractly in terms of nonphysical concepts.

It doesn't help when our current conventional spiritual models tend to lead us to incorrectly believe our spiritual part of life will not begin until our physical part of life is over. For example, some of the conventional spiritual models would lead us to believe we will go either to heaven or to hell after our physical life ends. Thus, we are bound to think the physical part of life is what life is all about as long as we are still alive on Earth. This is incorrect.

The new spiritual model is able to show how it is that the spiritual part of life and the physical part of life are going on at the same time while we are alive on Earth. After the physical part ends, the spiritual part simply continues to go on as a part of the spirit world forever. The new spiritual model indicates places like heaven and hell with all their numerous physical-like features are only concepts and not practices in the spirit world.

Spiritual guidance from the religious models is expressed in terms of physical concepts and this also keeps us focused on our physical part of life. Thus, such guidance is out of sync with the non-physical nature of the spiritual part of life even though the guidance is supposed to be based on the spiritual part of life. Therefore, the acceptance of such guidance has to rely on faith rather than on logic. A reliance on faith is much weaker than a reliance on logic.

We use our spiritual part all the time as we are going about with our daily lives. We look for and find our thoughts in the spiritual part of our lives and we express them in the physical part of our lives. When we invent something, we first explore and examine the various spiritual entities in the spiritual part of our lives to see if the invention has a chance of working before we proceed to physically construct the invention.

Such examples clearly indicate our spiritual part of life has to be going on the same time our physical part is going on and that both parts are interacting to make life on Earth able to function.

The new spiritual model indicates our creator does not need us to worship it. Instead our creator needs us to understand we have a partnership relationship with it, and we and our creator were meant to work together for the mutual benefit of both us and our creator. By having a partnership relationship means we also have the responsibility to make the partnership work.

The idea of our having such a responsibility has never been mentioned in any of the religious spiritual models. Instead worshipping was emphasized. But, worshipping would not be fulfilling our responsibility. Thus, we need to instead start emphasizing the responsibility we have.

The responsibility amounts to caring for our creator as well as receiving care from our creator.

4.4. Why Highly Religious People Could Struggle with Their Faiths

The article, "Highly Religious People Struggle the Most with Faith when They Experience Suffering, Study Finds", by Eric W. Dolan, March 22, 2023, PsyPoat, in Mental Health, Psychology of Religion, Social Psychology, Reference 10, was recently published.

At first glance, the finding in a study was surprising. We would tend to think highly religious people would struggle less while experiencing suffering.

But upon thinking about it further, there could be a logical reason behind the finding. It is mainly because our current religious models are not very complete and not very accurate. Thus, they are not able to provide very complete or very accurate spiritual guidance.

Therefore, a deeply religious person would be relying very strongly on spiritual guidance that is not very complete or very accurate. This means when such person is suffering, the spiritual guidance he or she strongly depends on would not be able to help him or her very much. Thus, he or she will struggle more that would someone who is not so deeply religious.

A person not deeply religious would have other resources to help him or her besides or instead of the spiritual guidance that religious models can provide.

Chapter Five

Attributes Regarding Our Purpose For Being Here on Earth

5.1. Attributes Regarding the Interactions Between the Two Parts of Life

The new spiritual model revealed the spirit world to be a living and growing thing. It also revealed a lot about how things work in the spirit world as indicated by the extensive list of features summarized in the Chapter Two.

In addition to the features, the model also provides a way to figure out the reason our universe was created and the primary purpose we humans are to fulfill while we are here on Earth. Numerous interesting secondary purposes were also revealed.

A living thing would naturally be a growing thing. We being a highly intelligent living thing, our growth would include spiritual advancements and technical advancements in addition to physical growth. Spiritual advancement is the knowledge we learned about how things work in the spirit world. Technical advancement is the knowledge we learned about how things work in our physical world.

This goes along with how we have a spiritual part of life in the spirit world and a physical part of life in our physical world with both parts going on at the same time while we are physically alive in our physical world. The spiritual part is really who we are. The physical part is temporary and exists to help the spirit world reestablish reasonable balance in its state of knowledge.

Both parts are involved with fulfilling the primary purpose for our being here on Earth. Thus, interactions will need to exist between the two parts to fulfill our purposes as well as to enable the physical part of life to exist. The various attributes of these interactions are as follows:

1. The interaction in which the spirit pilots the physical body has been covered earlier and will not be further discussed in this chapter. It is mentioned here only to be more complete in our listing of attributes of the interactions between the two parts of life.

2. Our spiritual advancements need to keep up with our technical advancements to provide spiritual guidance on how to apply or not to apply our technical advancements as new technical advancements are achieved. This topic has also been covered earlier and will not be further discussed in this chapter. Again, it is mentioned here only to be more complete in our listing of attributes of the interactions.

3. Compatibility between how we carry out our life and how things work in the spirit world is an important part of just about everything we do. It was briefly touched upon earlier, and we will see why it should a part of our lives in general as the presentation continues.

4. We must have free will in carrying out the interactions between the two parts of life. This was mentioned earlier and will be covered further in this chapter.

5. A natural partnership exists between the two parts of life. This

was mentioned earlier. It will be discussed in greater detail in this chapter to indicate how it is part of the new way of thinking that is emphasized in this book.

6. Our various secondary purposes for our being here on Earth were revealed by the new spiritual model and will be discussed in this chapter as well as in various other places as the presentation continues.

7. We should be striving for the equivalence of a "spiritual model of everything" much as how we have been striving for the equivalence of a "technical model of everything". We will see how this is also a part of the new way of thinking as the presentation continues.

The quality of human life on Earth depends on the quality of the interactions between the two parts of life. Right now the interactions are not working as well as they should be, and this is partly why we humans are making a mess on Earth.

The main reason is we have lost much of our awareness of the spiritual part of our lives. It is logical that in order for us to have good interactions between the two parts of our lives we have to be very aware of both parts of our lives.

5.2. Pursuing Our Primary Purpose in a Manner Compatible with the Spirit World

As explained earlier, the reason our universe was created was to help reestablish reasonable balance in the state of knowledge in the spirit world, and our primary purpose is to help our universe fulfill its reason for being created. The interactions mentioned in the preceding section of this chapter are a part of how we would fulfill our primary purpose.

In order for the interactions to effectively enable us to fulfill our primary purpose, the new pieces of knowledge we generate in our physical world has to be compatible with how things work in the spirit world.

Our spirit is a small portion of the spirit world. Our physical body exists in our physical world because our spirit enables it to exist. Thus, our body is in effect a physical representative of the spirit world. This means we are spiritually in the spirit world at the same time we are physically in our physical world.

Thus, in essence our physical body is to help the spirit world learn how to reestablish reasonable balance in its state of knowledge. This why the new pieces of knowledge we generate have to be compatible with how things work in the spirit world. It is so that the spirit world would be able to use the new pieces of knowledge to counterbalance the pieces of knowledge that are causing the imbalance.

By analogy, if we want to repair a complex machine, we need to know how the machine works in order to do what is necessary to repair it. The actions we take to repair it have to be compatible with how the machine works.

Up to now, we have not learned enough about how things work in the spirit world to know how to do a more complete and more accurate job of helping the spirit world reestablish reasonable balance.

More specifically, such actions as making wars, mistreating one another, and doing criminal acts of all kinds are not compatible with how things work in the spirit world. Instead of helping to reestablish reasonable balance, such actions are most likely causing additional imbalances to develop.

Such actions indicate we need to be able to form a whole lot more wisdom, and this means we need to achieve a whole lot more spiritual advancements. As explained earlier, wisdom is formed in the spiritual part of life where the knowledge of how things work in the spirit world is located.

5.3. We Need Free Will to Fulfill Our Primary Purpose, But Risks Exist

In order for highly intelligent living things to help reestablish reasonable balance for the spirit world, they must have the freedom to make choices. The spirit world would design and create a new universe that enables the kinds of experiences to be formed that would generate the kinds of new pieces of knowledge that could help restore reasonable balance.

But, the spirit world would not know the exact experiences that could be formed, only the kind of experiences that could be formed. Otherwise the spirit world would already have the pieces of knowledge that the new universe was created to generate, and thus it would not need to create the new universe. All this was explained in detail in Reference 1.

Therefore, it is up to the highly intelligent living things residing in the new universe to form the experiences and go through them to generate the kinds of new pieces of knowledge needed. This requires the living things to have the freedom to make choices.

However, the risk of giving highly intelligent living things this freedom is that the living things are likely to make bad choices as well as good choices. This is a risk the spirit world has to accept. It is because the spirit world could never be totally complete or absolutely perfect, and if the spirit world were totally complete and absolutely perfect it would not be living thing, as explained earlier.

So, we can see how if we say God is all knowing and all powerful we would not be able to figure out why our universe was created or what purpose we have for being here on Earth. This would leave mankind to aimlessly spend time doing destructive and meaningless things such as making wars, mistreating one another, and doing criminal acts of all kinds.

Thus, the reality of life is that mankind needs free will, but free will has its negative side as well as its positive side. This conundrum seems to be common in life. It is has to be a part of how things work for the origin of life. Otherwise the origin of life would not be able to originate life.

The various vicious circles mentioned earlier are the various forms of this conundrum. Thus, it is common to find living things doing things or having features to compensate for such conundrums.

For example, the spirit world would build safety factors and safety margins into various things it creates. This could be why we humans tend to also build safety factors and safety margins into a lot of the things we build. A lot of living things would have certain structural redundancies and/or certain regenerative and healing powers.

A possible example of a safety margin the spirit world build into our universe is that our universe is enormous to enable multiple planets to be hospitable for highly intelligent living thing so as to increase the chances that some planets will fulfill its primary purpose even if some planets would not. Unfortunately, Earth so far seems to be one of those planets that would not fulfill its primary purpose.

Also, more than likely, the spirit world would create different species of highly intelligent living things to reside on different hospitable planets. Again this is such that some species would succeed in fulfilling their primary purpose and some species would not. Again, unfortunately, we humans on Earth so far are not fulfilling our primary purpose.

Thus, we can expect the highly intelligent living things on other hospitable planets to be different from us humans. The fact that UFOs and UAPs seem so far to be very different from us humans could be evidence of this.

A possible safety margin the spirit world designed into us Humans on Earth is that we humans have a large number of races. The spirit world's intent was likely to provide us humans with a wide range of experiences and knowledge about life by means of a wide racial diversity to give us humans an advantage in being able to form the wisdom needed to figure out how to live our lives more constructively and more meaningfully.

In other words, racial diversity was meant to be a gift to mankind. Unfortunately, some of us humans are abusing this gift instead of valuing and embracing it and using it to our advantage. This is another one of the risks of giving us humans the freedom to make choices.

5.4. A Natural Partnership Is Formed between Every Living Thing and the Spirit World

When a living thing is called upon by the spirit world to reside in a new universe, that small portion of the spirit world that makes up the spirit of the living thing would enable the body of the living thing to exist in the new universe to become a representative of the spirit world in the new universe. As explained earlier, the spirit produces the body to exist in the universe by essentially thinking it to exist.

The living thing would then consist of two parts, a spirit that resides in the spirit world and a body that resides in the new universe. Because the spirit of the living thing is still a portion of the spirit world, a partnership is naturally formed between the living thing in the universe and the spirit world.

Accordingly, the living thing would then also have two parts to its life, a spiritual part that is going on in the spirit world and a part that is going on in the universe. Both parts would be going on at the same time and would be interacting with each other to make the body in the universe a living thing in that universe.

The partnership was meant to be for the benefit of both the living thing and the spirit world. However, such partnerships do not always end up being mutually beneficial. This is because the living thing has to have free will as explained in the preceding section of this chapter.

As a representative of the spirit world, the living thing ought to behave in a manner that is consistent and compatible with how things work in the spirit world. But, with free will, the living thing might not behave in such a manner, especially if the living thing happens to have lost much of its awareness of the spiritual part of life that is going on at the same time the part of life in the universe is going on.

This is what seems to have happened with us humans on Earth.

Accordingly, we were living mostly only our physical part of life. This means we have been trying to carry out life with only half of the knowledge regarding how life works, and we have been missing out on the wisdom that enables us to figure out how to live life more constructively and more meaningfully. Thus, we need to do the following:

1. Regain full awareness of the spiritual part of our lives and thus reenergize our pursuit of spiritual advancements.

2. Understand we have a natural partnership with the spirit world, and it is our responsibility to make the partnership be beneficial to both us and the spirit world.

5.5. Multiple Series of Universes Are Likely Created And Are still Being Created. This Produced Our Secondary Purposes

Reestablishing reasonable balance for the spirit world is likely to involve more than a single event and more than just one new universe. It is likely to be more like how when we go out to our garden do a specific task we would discover a lot of additional tasks waiting for our attention. Thus, we would end up spending the whole day in the garden instead of the short time we had planned.

The spirit world is extremely large such that at any point in time it is bound to have numerous issues needing to be rebalanced. This means many new universes are likely needed to address all the imbalances.

As explained earlier, everything in the spirit world is directly or indirectly a part of everything else in the spirit world. Consequently, reestablishing reasonable balance for one issue will likely change everything else in the spirit world to some degree ranging from negligible to major.

Some of these changes would cause additional issues to develop that then need more new universes to address. Thus, multiple intertwining series of universes could be constantly being created and are coexisting and functioning at the same time.

As the presentation continues we will discuss evidences that indicate multiple universes are likely to be coexisting.

When we humans finally get our act together and start focusing on fulfilling our primary purpose we are likely to discover we also have a lot of secondary purposes to fulfill as well. These would be produced by the changes that happen though out the spirit world as the various other highly intelligent living things in the various other new universes are fulfilling their primary purposes.

We might ask: how can the spirit world keep track of, and manage, all that is going on at the same time? The answer is at any point in time the spirit world possesses every piece of knowledge that exists at that point in time. And, it is these same pieces of knowledge that are making up all that is going on in the spirit world. Therefore, at any point in time the spirit world would automatically know everything that is going on and would be able to also manage all that is going on.

By comparison, each of us humans possesses only a relatively small portion of all the pieces of knowledge that go into creating our universe including all the living and nonliving things in it. Therefore, no one person is able to know everything that is going on in our universe, or even just within his or her local community.

The following evidences indicate numerous intertwining series of universes are continuing to be created and that multiple universes could be coexisting and functioning at the same time:

1. A series of universes (one series being created and another being created, another and another, etc.) could explain how some species of living things on Earth became more advanced than other species on Earth. It is that some species emerged earlier in a series earlier than others and thus have more time to evolve and advance.

We humans are the most advanced living things on Earth. Thus, humans likely emerged the earliest in a series of universes among all the species on Earth.

2. Humans have no fossils of a "missing link". Therefore, humans likely emerged on Earth through retracing. As explained earlier, retracing means emerging by going through an evolutionary path that has already been established in a past universe. Therefore, the emerging process went so quickly that no fossils of a missing link were left behind.

Conversely, no fossils being left behind means past universes must have existed. Thus, the evolutionary history of some highly advanced species such as us humans most likely spanned over multiple past universes. Some species being highly advanced while most other species are less advanced would indicate series of universes most likely have been created and are likely still being created.

3. Our dreams tend to be surrealistic. Dreams are produced as we sleep while our spirit is restoring our physical body from the wear and tear it received during the day. The restoration process consists of spiritual signals issued by all the pieces of knowledge making up our spirit to return our body back to its state of being before the wear and tear happened.

Our spirit is made of all the pieces of knowledge generated from experiences that took place in numerous past universes. Each universe is different and unique. Thus, the images from each universe would be different and unique. Therefore, the restoration process would involve spiritual signals that would produce images from pas universes. When all such images are combined, our dream images would be surrealistic.

Conversely, surrealistic dream images would indicate series of past universes are most likely to have been created and are most likely still being created.

4. According the SETI Institute, thousands of UFOs are sighted every year worldwide. The majority could be explained as being things such as weather balloons, airplanes, etc. that originated on Earth. A minority (some say roughly 20%) is unexplained, and these are the ones of interest in this discussion.

In this discussion, UFOs includes unexplained things that made crop circles. Crop circles would be among the various messages the UFOs are sending to us. In general such messages appear to be friendly and positive instead of hostile and negative.

I speculate that such UFOs and UAPs could be highly intelligent living things from coexisting universes visiting Earth to get us humans going at fulfilling our primary purpose for being on Earth. They appear in strange shapes because they are made of matter and energies different from the matter and energies making up our universe. Thus, our five major senses can sense them only partially instead of completely. This could be why they appear to us in strange shapes.

Conversely, their strange shapes could indicate coexisting universe are real and that they are functioning the same time our universe is functioning.

5. If highly intelligent living things in other universes are advanced enough to do universe travel, then they are likely to know how to make themselves at least partly sensed by our five major senses. Therefore, we would be able to sense them as UFOs and UAPs. Otherwise we would not be able to sense them at all. This way they are able to at least get they message conveyed to us

even if the messages are sort of vague at this time.

By being more advanced than we are, they most likely have fulfilled their primary purposes and are ready and waiting to start fulfilling their secondary purposes. But, the secondary purposes are to be jointly pursued with us humans, and we humans have to fulfill our primary purpose first. Thus, their visits to Earth are likely to persuade us humans to get our act together and fulfill our primary purpose.

6. The secondary purposes would be the various highly intelligent living things from various universes getting together after they have fulfilled their primary purpose to combine their results to come up with a more complete meaning for their efforts.

It is possible that the effort to reestablish reasonable balance is so complex that the effort needed to be divided into segments with each segment handled by a different universe. Therefore, once the individual segments are completed, the secondary purpose would be to combine the results from all the segments to thus complete the overall restoration process.

This would be analogous to how we humans would go about constructing anything that is complex such as a house, an automobile, a hospital, etc. Different companies would build different components making up the completed thing, and at some point in the construction process, these various companies would get together to be sure everything will fit together properly to form the completed complex thing.

7. Allergies could be an indication various different living things on Earth evolved in different past universes. By having evolved in different universes could result in some incompatibilities to occur between certain living things on Earth such as allergies.

Therefore, allergies could be another indication that multiple past universes had existed. Such multiple universes could be a part of a series of universes that is likely to still being created.

8. Our universe was designed to provide us humans with plenty of natural resources to enable us to achieve both spiritual advancements and technical advancements to where we would eventually be able to fulfill their primary purpose and then to go on to fulfill our secondary purposes.

We humans on Earth could have reach such levels of achievements if we had not wasted so much of our time, life, and natural resources making wars, mistreating one another, and doing criminal acts of all kinds for thousands of years.

One of the core problems we humans got trapped into is that for a long time we kept science and spirituality separate from each other as if one has nothing to do with the other. This was a huge mistake because the existence of life depends on both science and spirituality interacting together.

By keeping science and spirituality separated we kept ourselves from formulating spiritual models that could describe how things work in the spirit world as completely and as accurately as possible by basing the formulation processes on both spirituality and scientific and engineering logic.

Instead we settled for spiritual models such as the religious models that are devoid of scientific and engineering logic and that ended up relying on faith instead of logic for their acceptance. The models are correct regarding their spiritual guidance but are not complete and accurate enough to be very effective. Thus, we humans have not made much progress in fulfilling our primary purpose for thousands of years.

We might ask: if there are other highly intelligent living things elsewhere in our universe, could they have fulfilled their primary purpose? And if they did, why have they not try to contact us to get us to do the same? The answer is they might not have fulfilled their primary purpose either. Or, they might have fulfilled it but they have not been able to find a way to contact us, at least not yet.

They might have tried to contact us by going through the spirit world, but we humans have not been able to sense their messages because we have not been much in touch with our spiritual part of life. That would be yet another reason we need to regain our awareness of our spiritual part of life

As explained earlier in the book, universe travel could be easier than space travel. A coexisting universe could be occupying the same location as our universe is located, but living things in either universe are unable to sense the presence of the other universe because each has its different set of major senses, and each universe is made of a different set of materials and energies. Thus, the universes would not be interfering with each other even through they are located in the same location.

The UFOs and UAPs are definitely not here to take over Earth. If they were, they would have done so long ago. After all, they have obviously achieved a much higher level of advancements than we humans have. This means they carried out their lives more constructively and more meaningfully than we humans have such that they are much less likely to have made wars, mistreated one another, and done criminal acts of all kinds.

Just because we humans would invade and take over other nations doesn't mean other highly intelligent living things would invade and take over other universes. With few exceptions, the science fiction stories and movies we humans compose tend to have it all wrong about living things in other universes. Such stories and movies tend to keep us humans in a terrible mindset about life, and it is keeping us from living our lives more constructively and more meaningfully.

The topic of UFOs and UAPs is covered further in a later chapter as the presentation continues.

Chapter Six

The Spirit World Could Create Things Soon After Being Initiated

6.1. The Number of Spiritual Entities Increases Exponentially with Every New Piece of Knowledge Generated and Added to the Spirit World

The new spiritual model explains how the spirit world would grow rapidly with every new piece of knowledge generated and added to it. Therefore, the spirit world would be able to create things soon after its existence is initiated.

Let us start at the beginning when the spirit world consists of just one piece of knowledge. Then let us see how many new spiritual entities could be formed with each additional piece of knowledge.

Definition of symbols:

n	The number of pieces of knowledge in the spirit world.
TSn	The total number of spiritual entities formed for "n" pieces of knowledge.
m	The number equal to (n-1). For example, if "n = 6" then "m = 5".
TSm	The total number of spiritual entities for "m" pieces of knowledge.
Sx	The number of spiritual entities made of "x" number of pieces of knowledge.

The following determinations were done by manually constructing geometric figures and manually counting the numbers of spiritual entities of each possible size for "n = 1" to "n = 6". The smallest size spiritual entity consists of just two pieces of knowledge connected together with one connection of the first kind. The largest size spiritual entity for any given number of pieces of knowledge consists of all the pieces of knowledge all interconnected with connections of the first kind:

1. One piece of knowledge cannot form any spiritual entity:

$$n = 1 \quad TS1 = 0$$

   a. Zero with one piece of knowledge:

$$S1 = 0$$

2. Two pieces of knowledge can form one spiritual entity:

$$n = 2 \quad TS2 = 1$$

   a. One with two pieces of knowledge:

$$S2 = 1$$

3. Three pieces of knowledge can form four spiritual entities:

$$n = 3 \quad TS3 = 4$$

   a. Three with two pieces of knowledge:

$$S2 = 3$$

   b. One with three pieces of knowledge.

$$S3 = 1$$

4. Four pieces of knowledge can form eleven spiritual entities:

$$n = 4 \quad TS4 = 11$$

   a. Six with two pieces of knowledge:

$$S2 = 6$$

   b. Four with three pieces of knowledge:

$$S3 = 4$$

   c. One with four pieces of knowledge:

$$S4 = 1$$

5. Five pieces of knowledge can form twenty six spiritual entities:

$$n = 5 \quad TS5 = 26$$

   a. Ten with two pieces of knowledge:
   $$S2 = 10$$
   b. Ten with three pieces of knowledge:
   $$S3 = 10$$
   c. Five with four pieces of knowledge:
   $$S4 = 5$$
   d. One with five pieces of knowledge:
   $$S5 = 1$$

6. Six pieces of knowledge can form fifty seven spiritual entities:

$$n = 6 \quad TS6 = 57$$

   a. Fifteen with two pieces of knowledge:
   $$S2 = 15$$
   b. Twenty with three pieces of knowledge:
   $$S3 = 20$$
   c. Fifteen with four pieces of knowledge:
   $$S4 = 15$$
   d. Six with five pieces of knowledge:
   $$S5 = 6$$
   e. One with six pieces of knowledge:
   $$S6 = 1$$

These determinations provided the insight for deriving the following equation that can be used for doing such determinations for "n" greater than six:

$$TSn = (2 \times TSm) + (n - 1) \quad \text{or} \quad TSn = (2 \times TSm) + (m)$$

This logic behind this equation is as follows:

- Each spiritual entity that exists before a piece of knowledge was added would form a new spiritual entity consisting of the old spiritual entity plus one more piece of knowledge. Meanwhile the old spiritual entity would still exist. Therefore, the old spiritual entity and the new one formed from it would be two spiritual entities. This is how the number of spiritual entities would double with each added pieces of knowledge. This is expressed by the (2 x TSn) part of the equation.

- There were "m" number of single pieces of knowledge existing before a piece of knowledge was added. When a piece of knowledge was added, each of the "m" single pieces of knowledge would form a connection of the first kind with the added piece of knowledge. This would form "m" number of new spiritual entities consisting of two pieces of knowledge. This is expressed by the (n – 1) or the (m) part of the equation.

- Thus, when the two part of the equation are combined, the equation would indicate that the number of spiritual entities would slightly more than double with each added new piece of knowledge. The slightly more than doubling is expressed by the (n – 1) or the (m) part of the equation.

- This is how the total number of spiritual entities would essentially exponentially increase with every added piece of knowledge even though the equation is does not include an exponential expression.

The following examples confirm the equation as its results matched the manually generated results presented above.

$$TS1 = (2 \times 0) + 0 = 0 + 0 = 0$$
$$TS2 = (2 \times 0) + 1 = 0 + 1 = 1$$
$$TS3 = (2 \times 1) + 2 = 2 + 2 = 4$$
$$TS4 = (2 \times 4) + 3 = 8 + 3 = 11$$

$TS5 = (2 \times 11) + 4 = 22 + 4 = 26$

$TS6 = (2 \times 26) + 5 = 52 + 5 = 57$

The equation was then applied to determine following results for "n = 7" through "n = 20":

$TS7 = (2 \times 57) + 6 = 114 + 6 = 120$

$TS8 = (2 \times 120) + 7 = 240 + 7 = 247$

$TS9 = (2 \times 247) + 8 = 494 + 8 = 502$

$TS10 = (2 \times 592) + 9 = 1{,}184 + 9 = 1{,}193$

$TS11 = (2 \times 1{,}193) + 10 = 2{,}386 + 10 = 2{,}396$

$TS12 = (2 \times 2{,}396) + 11 = 4{,}792 + 11 = 4{,}803$

$TS13 = (2 \times 4{,}803) + 12 = 9{,}606 + 12 = 9{,}618$

$TS14 = (2 \times 9{,}618) + 13 = 19{,}236 + 13 = 19{,}249$

$TS15 = (2 \times 19{,}249) + 14 = 38{,}498 + 14 = 38{,}512$

$TS16 = (2 \times 38{,}512) + 15 = 77{,}024 + 15 = 77{,}039$

$TS17 = (2 \times 77{,}039) + 16 = 154{,}078 + 16 = 154{,}094$

$TS18 = (2 \times 154{,}094) + 17 = 308{,}188 + 17 = 308{,}205$

$TS19 = (2 \times 308{,}205) + 18 = 616{,}410 + 18 = 616{,}428$

$TS20 = (2 \times 616{,}428) + 19 = 1{,}232{,}856 + 19 = 1{,}232{,}875$

This indicates over a million spiritual entities could be formed with just 20 pieces of knowledge. The number of spiritual entities slightly more than doubles with every new piece of knowledge added. This slightly more than doubling was also shown to be the case in References 1 and 2 by using the line of logical reasoning presented without formulating an equation to explain it explicitly mathematical.

Reference 2 showed how just 30 pieces of knowledge would form 1,073,741,824 spiritual entities. This, it doesn't take long before the spirit world is able to create living and nonliving things after being initiated.

Going further in Reference 2, one million pieces of knowledge could form 499,999,500,000 spiritual entities. By now, today, the spirit world is bound to contain way more than one million pieces of knowledge. This is why there are plenty of spirits available to enable things such as the following to exist:

1. Lawns everywhere that are made of a countless number of grass plants.

2. A countless number of bacteria and/or viruses that cause pandemics.

3. Colonies of ants, termites, bees, etc could be found everywhere on Earth.

4. Hundreds of millions of people residing on Earth, and their number keeps increasing.

5. Countless number of celestial objects made up our particular physical universe.

6. Etc.

6.2. The Physicists' Model and the New Spiritual Model Agreed

As explained in the preceding section of this chapter the new spiritual model indicates the spirit world could start creating living and nonliving things soon after its existence is initiated.

The physicists' model presented in Reference 3 indicates the creator is able to create an extremely large number of things each time it fires up its mechanism to create things and that the mechanism could work soon after the creator was brought into existence. This means their model also indicates the creator could start creating things soon after the creator is brought into being.

This is one a several matches that was found between the new spiritual model and the physicists' model. Other matches are described as the presentation continues.

The particular match describe at this time is the most significant one since it is about how the creator managed to create a countless number of things to exist on Earth. Having a countless number of a countless variety of different things around us is a part of our everyday experiences and observations. This phenomenon enables eco systems that are hospitable for living things to be formed and the possibility of providing food for all the living things that exist at any given point in time.

The fact that the two models matched each other in several different ways tends to provide confirmation for both models. This is particularly significant since their formulation processes are so different.

More specifically, the physicists took spirituality into consideration in formulating their scientific model. References 1 and 2 took scientific and engineering logic into consideration in the formulation of the new spiritual model. In other words, each started from the opposite end of the same spectrum.

Chapter Seven

Dark Energy & Dark Matter, Antimatter, Boundaries of Universes, Background Microwave Radiation, Gravitation Waves in Space, and How Things Are Brought into Existence without Creating Antimatter

7.1. The New Spiritual Model Offers Possible Answers to Unanswered Questions about Our Universe

Creating something out of nothing such as with the Big Bang would create matter and antimatter in equal amounts. But, no antimatter has been found so far in our universe except when produced in a laboratory. Therefore, our universe might have been created by some other means than by the Big Bang.

Our universe is said to consist of roughly 68% dark energy, about 27% dark matter, and approximately 5% normal matter, Reference 11. How this came about has not been clearly explained so far.

Also not clearly explained is why background microwave radiation exists? The radiation was assumed to be leftover from the Big Bang. But, what caused the Big Bang, and how it could produce the background radiation are unanswered questions.

The new spiritual model indicates the space that exists as part of our universe is a physical thing instead of a nonphysical thing as we might assume. Einstein's theory of relativity could explain how gravitational waves could be generated in space by massive black holes. This would also suggest space is a physical thing.

Also an unanswered question is what produced the gravity that is a part of our universe?

The new spiritual model is able to provide plausible logical explanations for the topics mentioned, as presented in the rest of this chapter.

7.2. Thinking Something to Exist Is Different from Creating Something Out of Nothing

The new spiritual model revealed a way the spirit world could bring living and nonliving things into existence in our universe that we humans have not thought about before. This even includes how the spirit world could bring our entire universe into existence.

This newly discovered way was revealed by the new spiritual model largely because the model could explain how things work in the spirit world more completely and more accurately than could any spiritual model that exists up to now.

The new spiritual model took science into consideration and it also used nonphysical concepts in the modeling process. This made the model more compatible with the nonphysical nature of the spirit world than any spiritual model that exits up to now. This enabled the new spiritual model to achieve be more complete and more accurate modeling of how things work in the spirit world.

As indicated earlier, all spiritual entities are structurally the same as spiritual entities that are thoughts. This includes all spirits of living and nonliving things since all spirits are spiritual entities.

This means every living or nonliving thing in our universe and our universe itself, exist because its spirit is essentially thinking them to exist. It is unconventional but is logical based on the new spiritual model and the structural makeup of all spiritual entities.

Thinking things to exist is not the same as creating something out of nothing. Therefore, it would not produce antimatter.

By analogy, it is like how the world of a story is made to exist by the author thinking it to exist. In the process of thinking the world of the story to exist, the author would not be thinking an anti-version of that world to exist.

The fact that no antimatter is found in our universe and no anti-version of our universe is seen tends to confirm the concept that our universe and everything in it exists because the spirit world is essentially thinking them to exist. This in turn tends to confirm the validity of the new spiritual model since the described concept was revealed by the new spiritual model.

On the other hand we might speculate the Big Bang somehow separated all the antimatter from all the matter and blew all the antimatter extremely far away to form an antimatter universe far away. The antimatter universe would be so far away we have not yet been able to see it even with our most powerful telescopes.

But, until we are able to see such an antimatter universe, it would be valid to consider our universe as having been brought into being by the spirit world essentially thinking it to exist.

In order for the spirit world to be able to create a universe as complex and ours and also living things as complex as those residing on Earth and most likely also elsewhere in our universe, the spirit world must have had a lot of experience creating things. This means a lot of past universes must have been created before ours was created.

The possible existence of past universes could explain a lot about things and phenomena we have been able to observe but have not been able to explain up to now. Explanations for some of such observations are presented as the presentation continues.

7.3. The Space in Our Universe Is A Physical Thing

Our universe is a physical universe. This means every part of it is physical. It is not a mixture of physical parts and nonphysical parts. This means even the space that is a part of our universe is physical. After all, before our universe came into being, there was only nothingness, and the nothingness would not be the space that is now a part of our universe.

In order for the space within our universe to be a physical thing it has to be made of something physical. I would speculate that space is made of yet to be discovered very tiny quantum particles that are transparent much like how clear glass and certain plastics are transparent.

These quantum particles could also be so much smaller than electrons, protons and neutrons they could even make up the space between the electrons and the nucleus that is within atoms of the atomic elements without interfering with the actions and functions of the atomic elements.

Thus, larger typical physical things can move through outer space without encountering resistance, and conversely outer space would not be disturbed by the motions of larger typical physical things. This could lead some people into thinking the space within our universe is nothingness. But, it is not nothingness. It has to be a physical thing in order to be a part of our physical universe.

The tiny quantum particles are so tiny they are yet to be detected by scientists. They would also be partly matter and partly energy just as any kind of quantum particles would be. They have to be partly energy in order to keep them from clumping together to form a liquid or a solid. It is like how when ice has enough energy it would become water, and when water has enough energy it would become water vapor. Therefore, right now the tiny quantum particles have enough energy to be like a gas to form the space in our universe.

The energy part of them is then the microwave radiation detected in space. Thus, the microwave radiation would not be residual radiation left over from the Big Bang. This means, according to this discussion, the Big Bang most likely never happened.

7.4. The Tiny Quantum Particles Could Also Be Making Up Gravity

The question of what is making up gravity is still not firmly answered. Numerous articles have been written that indicate this to be the case. The latest such article published in mid-year 2023 is "A New Study Appears to Stunningly Contradict Newton and Einstein's Theories of Gravity", by Darren Orf, Popular Mechanics, August 14, 2023, Reference 12.

In view of gravity being not firmly determined as to what is making it up, another possibility is that the tiny quantum particles introduced in the preceding section of this chapter as possibly making up the space in our universe could also be possibly making up the gravity in our universe.

The following are several descriptions of possible behaviors of the tiny quantum particles that are consistent among each description and consistent with observations.

1. The tiny quantum particles could also be a part of every physical thing in our universe by making up the tiny spaces within every atom that makes up all physical things. The tiny quantum particles could thus produce a gravitational attractive force that exists between each physical thing and every other physical thing in our universe.

2. This same gravitational attractive force could be part of how atoms are formed. Besides the attractive force that exists between the negative charge of electrons and the positive charge of protons, the gravitational attractive force produced by tiny quantum particles that make up the tiny spaces within atoms could also be a part of enabling electrons, protons and neutrons to stay together in a manner that could form the various kinds of atoms.

3. Black holes are made of subatomic particles other than these tiny quantum particles. The subatomic particles would include electrons, protons, neutrons, and other subatomic particles scientists have identified so far. These subatomic particles are packed so tightly within a black hole that the tiny quantum particles making up space are squeeze out of the black hole. Thus, no space exists within a black hole.

4. The tiny quantum particles squeezed out would be concentrated around the outside immediately next to the black hole such that their density there would be higher there than anywhere else in space. This could explain how space is warped immediately next to a black hole.

5. The larger the black hole the more tightly packed would be the

tiny quantum particle immediately next to the black hole, and therefore the higher would be the gravitational force they would produce there. Thus, the space surrounding the black hole would naturally be more warped.

In this sense, the space formed by the tiny quantum particles would act sort of like plastic foam such that when we push a solid object into the foam, the density of the compressed foam surrounding immediately next to the solid object would be more dense than anywhere else in the foam.

6. This also means these tiny quantum particles cannot pass through black holes as black hole move around in space. Therefore, when black holes are moving around in space they are pushing these tiny quantum particles around.

Thus, when black holes are moving fast enough through space they could generate waves in space. And since these tiny quantum particles also make up gravity, the waves generated would also be gravitational waves.

An example of such an occurrence is described in Reference 13, and this confirms space is a physical thing. As described in Reference 13, a highly sensitive instrument was constructed to detect the gravitational waves generated by two massive black holes moving near the speed of light and will be colliding together.

7. These tiny quantum particles could also limits how fast light and other kinds of energy waves could travel through space. It is like how sound waves could travel through solids and various liquids faster than they could travel through space. Similarly, light would travel through various transparent liquids and solids at a speed that is different from the speed through space.

8. Therefore, one thing we could say about the tiny quantum particles is that they certainly behave in ways that are different from the usual physical substances we are used to seeing, touching, and using in our daily lives. This seems to be true of all quantum particles, and that is why Newtonian physics theories do not apply for quantum particles. They require quantum mechanics or quantum physics theories to describe their behavior.

The tiny quantum particles that make up space and gravity might require yet a different kind of quantum mechanics or quantum physics theories to apply for them. This is because they are much smaller than the usual kinds of quantum particles and their behavior is many ways different from the way the usual kinds of quantum particles behave.

The descriptions of several possible kinds of behavior of the tiny quantum particles as presented in this section of this chapter turn out to be all consistent with each other and are also consistent with observations. Therefore, the tiny quantum particles could indeed make up space and the gravity that are a part of our particular physical universe.

7.5. Space in Our Universe Could Be Made of Dark Energy & Dark Matter

As explained earlier in this chapter, the space within our universe could be made of tiny quantum particles that are so small they are yet to be detected by scientists. As explained earlier in this book how quantum particles would appear as energy waves when observed one way and as matter when observed in another way.

The new spiritual model indicates that this is because the tiny quantum particles that appear to make up space and gravity reside at the interface between the spirit world and our physical world.

Thus, a tiny quantum particle would have two sides, one side facing the spirit world and the opposite side facing our physical world. The side facing the spirit world would be made of spiritual energy issued from the spirit world. The side facing our physical world would be made of physical matter.

Thus, if we happened to observe the tiny quantum particles on their side facing the spirit world, they would appear as being what we call dark energy. And, if we happened to observe them on the side facing our physical world, they would appear as what we call dark matter.

As mentioned earlier, our universe is said to consist of roughly 68% dark energy, about 27% dark matter, and approximately 5% normal matter, Reference 11. How this came about has not been clearly explained so far. However, a possible explanation is introduced here and is as follows.

1. The tiny quantum particles might have a configuration similar to a solid hemisphere. The convex side faces the spirit world and would appear as energy and the flat side faces our physical world and would appear as matter.

2. The convex side has the larger area and therefore it is possible to see the convex side 68% of the time. The flat side has the smaller area and therefore it is possible to see the flat side 27% of the time.

3. Mathematically, the area of the convex side is 2.0 times the area of the flat side. This is somewhat close to how 68% is 2.52 times as large as 27%.

4. Being hemisphere in shape, the convex side could simply be seen more often than the flat side in addition to its area being 2.0

times the area of the flat side. The sum total of these two attributes could result in the convex side being seen 2.52 times more often than the flat side.

This could explain why dark energy is observed to exist 68% of the time and dark matter 27% of the time because 68% is 2.52 times 27%.

Volume-wise our universe is practically made entirely of space with solid and liquid matter making up only a tiny fraction of the total volume. Therefore, even though the density of dark energy and dark matter making up space is very thin, the total amount of them making up our physical universe could very well add up to 95% (68% + 27% = 95%) of our physical universe leaving only 5% consisting of solid, liquid, and gaseous physical matter.

All this is assuming space is made of dark energy and dark matter and that dark energy and dark matter are simply the two different attributes of the tiny quantum particles that make up space and gravity as I am proposing. The numbers and the mathematics tend to suppose this proposition.

7.6. The Background Microwave Radiation Could Be an Attribute of Dark Energy & Dark Matter

The background microwave radiation is thought to be leftover from the Big Bang. But, according to the new spiritual model, the radiation could instead be an attribute of the tiny quantum particles that appear to make up space and gravity.

As explained earlier in this chapter, it is the energy the tiny quantum particles must have to keep them from clumping together to form a liquid or a solid. By analogy, it is like how steam must have enough energy to keep from becoming liquid water and how liquid water must have energy to keep from becoming ice.

Also as explained earlier in this chapter, the tiny quantum particles could also be the dark energy and dark matter that are detected to exist in space. Thus, according to the new spiritual model, the dark energy and dark matter are what the space in our universe is made of.

The net result is that the background microwave radiation could be a natural attribute of dark energy and dark matter. It is the energy the tiny quantum particles making up space, gravity, and dark energy and dark matter have that is keeping the tiny quantum particles from becoming a liquid or a solid.

7.7. Two Ways to Release Nuclear Energy:
Fission: Splitting Very Heavy Atoms
Fusion: Fusing Very Light Atoms

The concepts introduced in this section of this chapter are pure speculations. They are not confirmed with observations or are backed up by established theories. They are things to think about that are "outside the box" simply for the fun of it.

The subjects addressed throughout this chapter are among ones with unanswered questions regarding our universe. Unconventional concepts associated with the new spiritual model were introduced in this chapter to produce unconventional but plausible answers to the unanswered questions.

The concept introduced in this section of this chapter is particularly unconventional. It is an unconventional extension of the unconventional concepts introduced in the earlier sections of this chapter. It is able to provide a possibly reason why two reactions that are essentially opposites could release nuclear energy that is in addition to the nuclear energy that is produced according to established theory.

Thus, the total release of nuclear energy might consist of two sources. The main source would be as explained by established conventional theories. The second source which is probably much smaller is speculated to take place as explained in this section of this chapter.

The main release of nuclear energy can be produced by either of following two kinds of nuclear reactions. They are essentially opposite reactions involving atoms at opposite ends of the spectrum of atomic elements.

> 1. **Fission:** Fission is the splitting of a heavy atom into to two lighter atoms. For example, when either a plutonium or an uranium atom, both are very heavy atomic elements, is split apart to form two lighter atoms, a huge amount of nuclear energy is released,

> 2. **Fusion:** Fusion is the combining of two light atoms to form one heavier atom. For example, when two hydrogen atoms, the lightest atomic element, are combined to form one helium atom, a huge amount of nuclear energy is released.

While the two reactions are essentially opposites, a commonality exists for the two reactions. It is a net reduction in the space that exists within atoms is produced by both reactions. More specifically, some of the space within atoms is released in both reactions.

If this space is made of the tiny quantum particles that are a major topic of discussion in this chapter, the question is, could this release of such space also produce some release of nuclear energy?

Such release of space within atoms could be shown to be true for both kinds of reaction as follows. The volume of a sphere is given by the expression: [(4.189) x (the radius cubed)] that could be found in any standard mathematical handbook containing equations for determining the volume of various three-dimensional shapes.

1. **Fission:** In fission, the space within a very heavy atom is approximately given by the above expression where the radius is the radius of the atom.

When the atom is split into two lighter and smaller atoms, the radius of each of the smaller atoms could be assumed to range from 0.5 to 0.75 of the radius of the original heavy atom. Then the combined space within the two smaller atoms would range from 0.25 to 0.84 of the space that was within the original heavy atom.

Thus, a release of space within atoms would range from 0.16 to 0.75 of the space that was within the original heavy atom.

2. **Fusion:** In fusion, the radius of a helium atom is approximately the same as the radius of a hydrogen atom. Therefore, combining two hydrogen atoms to form one helium atom would mean a release of space within atoms of 0.5 of the combined space that was within the two original hydrogen atoms.

Since a hydrogen atom is much smaller than a plutonium or uranium atom, a release of 0.5 for the combined total of two hydrogen atoms could be within the same order of magnitude as a release of 0.16 to 0.75 for a plutonium or uranium atom.

Thus, each of the two kinds of nuclear reactions would produce a release of space within atoms, and the amount of release by each kind of reaction could be comparable.

If the tiny quantum particles are able to make up space, gravity, dark energy and dark matter, and the background microwave radiation, then it would seem possible they could have a significant role in the formation of atoms for all atomic elements by forming the space that exists within the atoms of every atomic element.

For example, they could provide the forces that determine how many electrons can occupy each orbit that goes around the nucleus of an atom.

Each atom of each atomic element must have required a lot of energy to be produced since heavier elements are produced from fusing lighter elements together in exploding stars and other massive events that take place in outer space.

Therefore, any changes to any of the atoms could result in the release of some of the energy that went into forming the atoms in the first place. And such changes could involve changes in the nucleus of atoms and also changes in the space within atoms.

This suggests that any release of space within atoms could involve some amount of release of nuclear energy. If this is true, then the release of nuclear energy in both of the two kinds of nuclear reactions would consist of two parts, a main part and a secondary part that comes from the release of space that is within atoms.

7.8. Universes Have Boundaries

As explained earlier in this chapter, the space that is a part of our universe has to be a physical thing and is likely made of tiny quantum particles. Since nothingness was there before our universe was created, there has to be a boundary separating our universe from the nothingness that surrounds our universe.

According to the new spiritual model, our universe or any other universe could not possibly extend out forever such there would be truly an infinite number of objects making up a universe that is infinitely large. This is because the new spiritual model indicates the spirit world is made up of a finite but ever increasing number of existing pieces of knowledge. This means only a finite but ever increasing number of spirits of things could be formed at any given point in time. Thus, an infinite number of things could never exist.

But, we might ask: what about as explained in Reference 1 that the spirit of elemental building materials can enable a limitless number of such building materials to exist. For example, only one spirit for electrons would enable a limitless number of electrons to exist. The same goes for protons, neutrons, subatomic particles, atoms of every elemental atomic chart, and various kinds of molecules such as water molecule.

The answer is while all this is true it is not practical for the spirit world to enable an infinite number of anything to exist. After all, the spirit world has to be aware and conscious of everything in order to be able to maintain reasonable balance within its state of knowledge.

Thus, on a realistic sense the spirit world is never going to be infinitely large and therefore could never be able to be aware and conscious of an infinity number of anything. Therefore, while theoretically the spirit world could enable an infinitely number of some things to exist, it would be suicidal for it to enable an infinitely number of anything to exist.

Also, if the spirit world were to decide to enable an infinite number of something to exist, it would be spending time nonstop enabling an infinite number of that something to exist because it could never keep tract of an infinite number of them. Thus, the spirit world would not have time to do anything else. Again, that would be suicidal.

Therefore, our universe, as with any other universe, has to be finite, and thus, as with any other universe, our universe would have a boundary.

As explained earlier, the tiny quantum particles that are likely making up space is also likely to make up gravity. This means these tiny quantum particles could be holding everything making up our universe together to form a universe that has a finite size and thus has a boundary.

A boundary would form because physical space and nothingness are not likely to mix.

By analogy, this would like how oil and water do not mix. A glob of oil in water would stay a finite size and would have a boundary between the oil and the water. This is because beyond the glob of oil is nothingness in terms of there would be no oil elsewhere in the rest of the water.

This means our universe could be like an infinitely flexible glob. The major different from a glob of oil is that our universe is a living thing whereas a glob of oil is not a living thing. Therefore, the spirit of our universe could reshape our universe in any way it sees fit.

This means when highly intelligent living things in our universe learn how to do space travel in a practical manner, and they space travel close to the boundary of our universe, the spirit of our universe could keep reshaping the boundary such that the living things will always have more space ahead of them to travel into. This could give them the illusion that our universe has no boundary and is infinitely large.

We might ask: why would the spirit of our universe do this since it would give living things the illusion that our universe has no boundary? The answer is it is only a speculation that the spirit of our universe might do this. The possible answer is the spirit of our universe might be protective of any highly intelligent living thing from any dangers that could result from being too close to the boundary of our universe.

It is also possible that space travelers nearing the boundary of our universe might try to send a small projectile through the boundary into the nothingness to see what could happen. Then the most likely scenario would be the boundary would be pushed out by the projectile and would then pull the projectile back into the universe. This is because the nothingness on the other side of the boundary has no space for the projectile to go into.

Under this scenario, the boundary would behave like an infinitely flexible wall that nothing physical could go through. The volume inside the boundary would remain unchanged while the boundary is being reshaped. Therefore, the interesting thing to observe would be how the celestial objects within our universe might be moved around a bit due to any reshaping of our universe.

However, the universe is so large that such movements would likely hardly be noticeable because such reshaping of the boundary would happen very locally and would be very small compared with the size of our universe.

7.9. Coexisting Universes Can Be Located On Top of Each Other

All universes, physical and nonphysical, would have a boundary, based on the discussion so far, regardless of their dimensionality or the physical or nonphysical substances and energies they are made of. This is because outside of their boundaries would be nothingness. Thus, there has to be a boundary to separate anything physical or nonphysical from the nothingness in every case.

It is more obvious physical universes would have a boundary. But, nonphysical universes would also have a boundary. Their nonphysical substances and energies are not nothingness.

What this means is all coexisting universes regardless of whether they are physical or nonphysical and regardless of their dimensionality and the substances and energies they are made of could coexist "on top of one another" and occupying the same location. This is because each universe is different and unique, and each set of physical or nonphysical substances and energies would not interfere with any other set of physical or nonphysical substances and energies.

Also, the major senses of living things in one universe would not be able to sense the substances and energies of another universe, because their major senses are tuned only to the substances and energies of their own universe. Thus, multiple universes could coexist in the same location, and living things in one universe would not know that other universes are located right on top of their universe.

This could make doing "universe travel" among coexisting universes much easier than doing "space travel" within any universe. This is because every coexisting universe would be "right next door with" every other coexisting universe.

Universe travel would be performed by going through the spirit world. This enables the universe travelers to gain access to various possible ways to become able to at least partially sense things in the universe they are traveling into. This would also render the universe travelers to be at least partially sensed by the living things residing in the universe the travelers have traveled into.

This could explain how UFOs and UAPs could appear in the sky of Earth and how crop circles could be formed in the fields on Earth. The UFOs and UAPs would appear as strange shapes to us humans because we are able to sense them only partially. More about UFOs and UAPs is covered later in this book.

All the coexisting universes being "right next door with" every other coexisting universe might be by design for a reason. As mentioned earlier, highly intelligent living things have secondary purposes. The universes are to interact with each other to be sure what each is doing could be consistent with each other and be compatible with how things work in the spirit world when all the results are combined together.

Therefore, by having all coexisting universes being next to each other would enable such consistencies to be more easily periodically checked and any necessary adjustments made. This would make more certain that consistency will be maintained. Such consistencies are important because, as explained earlier, the new pieces of knowledge would be effective in reestablishing reasonable balance only if they are generated by experiences that are compatible with how things work in the spirit world.

The UFOs and UAPs visiting Earth might be here to interact with us humans for such reasons. But, they found us not doing well in fulfilling our primary purpose. Therefore, their first reason for visiting us might be to try getting us to work toward fulfilling our primary purpose. They try to do this through messages they are conveying to us the best they can.

We should study their actions and their crop circles to identify the messages that are embedded in their actions. One such message might be saying "we are here and we are waiting for you to interact with us in a positive manner." After all, UFOs and UAPs have never been hostile. As mentioned earlier, just because we humans would invade another country to take it over doesn't mean other highly intelligent living things in other universes would behave so destructively, unwisely, and meaninglessly.

By deciphering why UFOs and UAPs are visiting us could go a long ways to inspire us humans to think longer term and longer range, and thus we would be more motivated to behave overall better and wiser. We would thus be more likely to live our lives more constructively and more meaningfully.

Chapter Eight

The Source of Consciousness and The Development of The Ability to Form DNA Molecules

As explained very early in this book, physicists formulated models that came close to bridging the gap between spirituality and science. They were unable to model: **(1)** how consciousness came into being and **(2)** how DNA molecules could be formed.

The new spiritual model formulated in References 1 and 2 was able to include both of these topics in its formulation process. This was accomplished by extending the new spiritual model to include the origin of life as explained in detail in Reference 2 and briefly explained as follows:

1. **Consciousness:** In extending the new spiritual model to include the origin of life, consciousness was one of the four qualities (4Qs) that make up the origin of life. The other three qualities are intelligence, curiosity, and wisdom. Their combined attributes enables life to exist as explained in Reference 2.

The 4Qs would exists before any pieces of knowledge exists and thus also before the spirit world exists. To be the origin of life, the 4Qs has to be spiritually a living thing, which means it has to be able to spiritually grow and/or to spiritually learn.

A living thing is a growing thing. Learning would be a form of growth.

By being conscious, intelligent, curious, and wise, it is naturally able to form and go through experiences which would then generate new pieces of knowledge. Thus, the 4Qs is naturally able to learn and therefore also grow and is therefore a living thing.

The first pieces of knowledge the 4Qs generates would naturally initiate the existence of the spirit world. The spirit world would be a living thing since it will be growing and learning as additional pieces of knowledge are generated by the 4Qs that would be added to it.

The spirit world would keep growing and learning in this manner and will eventually be large enough to be self sustaining. It would then be able to create living and nonliving things on its own that would go through experiences and generate new pieces of knowledge that would also be added to the spirit world.

Thus, its rate of growth and learning could increase in an exponential manner.

In order for any thing to be a living thing, it would have to have one or more of the four qualities making up the origin of life. It is logical that all living things must have at least consciousness. Plants would be as example of living things that have at least consciousness. For example, a plant is conscious of where there is sun light and where there is water.

Thus, according to the new spiritual model, consciousness came about because it is part of the origin of life

2. **The Ability to Form DNA Molecules:** As explained earlier,

everything that exists as being a part of our physical world is created by the spirit world. Since DNA molecules are a crucial part of how living things in our physical world are created, the ability to form DNA molecule has to be a part of the creative ability of the spirit world.

As with any ability, the ability to form DNA molecules has to be gradually learned. After all, the spirit world itself gradually grows by having new pieces of knowledge added to it one piece at a time. Thus, the spirit world learned how create anything, including how to form DNA molecules, by having new pieces of knowledge added to it one piece at a time.

This means our universe must have been preceded by many previous universes by which the spirit world learned how to create a whole bunch of new things before it created our universe.

Each universe is different and unique as explained earlier. Therefore, there are likely many different ways to constructing living things to be living things depending on the universe they reside in. Thus, there must have been many earlier more primitive ways to construct living things that eventually enable the spirit world to learn how to construct DNA molecules.

Thus, according to the new spiritual model, the spirit world's ability to form DNA molecules must have been gradually learned one new piece of knowledge at a time. The learning process must have involved a sequence of universes with each universe being more advanced than the one preceding it.

Conversely, the ability of the spirit world to know how to create DNA molecules could be an indication that sequences of universes have been created in the past as are still being created now.

Chapter Nine

Creatures that Retained Their Awareness of Their Spiritual Part of Life

Animals and various other creatures in general appear to have retained their awareness of the spiritual part of life better than we humans are able to retain. Rupert Sheldrake in Reference 8 presented numerous examples of how some dogs, cats, horses, cattle, birds, etc. are able to do things that can be done because, in my opinion, they retained their awareness of the spiritual part of life.

Examples of what they could are: **(1)** telepathically read their owner's mind, **(2)** find their way to certain places on their own, **(3)** find their owner who is now temporarily located in an unfamiliar place in another country, **(4)** locate their deceased owner's gravesite, **(5)** know danger lies ahead and thus alert its owner to go no further, **(6)** learn about something through telepathy what other members of their species have learned, etc.

According to the new spiritual model, in order for animals and other creatures to be able to do such extrasensory things, they have to know how things work in the spirit world. This is because such extrasensory actions are performed by going through the spirit world to do them.

The knowledge regarding how things work in the spirit world and the wisdom that knowledge can form are found in the spiritual part of life, as explained earlier in this book.

Besides the extrasensory things animals and various creatures are able to do, their behavior also indicates they are able to gain access to the wisdom that can be found in the spiritual part of life. More specifically, they do not make wars, mistreat one another, and do criminal acts of all kinds. While their lives tend to be much simpler than how we humans live our lives, they live their lives constructively and meaningfully.

Thus, while we humans are way ahead of animals and creatures in terms of technical advancements, they appear to be way ahead of us humans in terms of spiritual advancements. This suggests we might be able to learn a lot from them regarding our pursuit of spiritual advancements.

Rupert Sheldrake's interpretation is that these animals and creatures are in tune with a morphic field that I would envision as providing them with a certain connection with the world that is in addition to the connection our human five major senses could provide us.

In my mind, the morphic field would be like how the oneness in the spirit world would is naturally formed in the spirit world because everything in the spirit world is naturally directly or indirectly a part of everything else.

The connections provided by the oneness are available for us humans to tune into, and it would give us extrasensory abilities too and would also help give us the wisdom to behave better and wiser, much like how animals and various creatures behave. Again, they do not make wars, mistreat one another, and do criminal acts of all kinds.

The animals and various creatures being in tune with the morphic field, or the oneness in the spirit world, would mean they also have an extrasensory connection with humans. This could explain how a lot of animals and creature could be domesticated by humans.

According to the new spiritual model, the connection among animals and creatures and humans enables certain highly intelligent creatures such as dolphins and whales to develop closeness with humans without being owned by humans. For example, one species of whales even works with fishermen to capture fish for the fishermen and for themselves, Reference 1, page 278.

A conclusion I reached from all such examples is that they reinforce the idea that every living thing in every universe has a spiritual part of live as well as part of life that pertains to the universe in which the living thing resides.

In the case of our physical universe, the reason animals and creatures retained their awareness of the spiritual part of life better than we humans have retained is because animals and creatures rely much more on the spirit world to do part of their thinking for them. We humans do almost all of our thinking ourselves and therefore we are naturally less close with the spirit world in this sense. Thus, we humans have a greater chance of forgetting about the spiritual part of life.

Chapter Ten

Telepathy and Various Extrasensory Abilities Should Be Considered Natural and Normal

10.1. Doing Things by Going Through The Spirit World

Being able to do things by going through the spirit world is largely made possible by the oneness that pervades the spirit world. As explained earlier in this book and in detail in Reference 1, the oneness was formed because the spirit world is made up of spiritual entities, and every spiritual entity is directly or indirectly a part of every other spiritual entity.

This in turn came about because only one copy of every existing piece of knowledge is kept. This is because, as explained in Reference 1, more than one copy and any existing piece of knowledge would not increase the state of knowledge for the spirit world. Thus, duplicates would merge and only one copy would remain.

This is an example of how scientific and engineering logic was applied throughout the formulation process of the new spiritual model as presented in References 1 and 2.

Much of the things we humans are able to do by going through the spirit world are considered paranormal or extrasensory. We are so used to how things work in our physical part of life, how things work in our spiritual part of life would seem abnormal to us. This is mainly because in our physical part of life everything is physically separate whereas in our spiritual part of life everything is directly or indirectly a part of everything else, as explained in Reference 1.

Manifestations of how things are directly or indirectly a part of everything else are all around us in our physical world. For example, every person shares over 90% of his or her genes with every other person. We humans even share some of our genes with other living things. Another example is every person has the same basic physical structural design, the same set of major senses, more or less the same consciousness and intelligence, etc.

More examples are that animals and creatures all reproduce with eggs and sperms. All plants have leaves or needles and they all reproduce by forming flowers and seeds. All birds have feathers, wings, a beak, etc. Etc.

When things happen in an interconnected ways in our physical world, it would be so foreign to us we would consider them paranormal or extrasensory. But, by thinking that way, we tend to not consider such interconnections as being a natural and spiritual part of life. We tend to ignore them or are fearful of them. This tends to keep us away from the spiritual part of life.

Based on the discussion in this section of this chapter, the following conclusions could be made. Such conclusions could help to open our minds to a new way of thinking and a new way of living our lives.

    1. If animals and creatures are able to do things by going through the spirit world in a natural way, then doing things by going through the spirit world must be a natural phenomenon. Thus, we should not ignore it or be fearful of it.

2. All living things at the time they were created are likely naturally able to do things by going through the spirit world. As they evolve, some retained this ability better than others. We humans are among the ones that did not retain this ability very much.

3. If we humans had this ability at one time, we should be able to regain it. We need to stop consider this ability as being extrasensory or paranormal. Instead we need to consider it as natural and normal and thus explore ways to regain it.

We tend to think having this ability is unnatural because we have forgotten we have a spiritual part of life going on the same time our physical part of life is going on. Therefore, let's take a look at some of the current paranormal and extrasensory things that have been shown to function in the spirit world earlier or in References 1 and 2 and are thus a natural part of the spiritual part of life.

Such paranormal or extrasensory things could be considered to be of three types. Examples of each type are covered in the next three sections of this chapter. Some examples are taken from Rupert Sheldrake's book, Reference 8:

10.2. Activities that Take Place in the Spirit World That Can Produce a Response in Our Physical World

Activities can take place in the spirit world that we in our physical world could sense and influence what we think and/or what we do. The following are examples.

1. **Seeing and reading auras:** According to the new spiritual model aura are produced by spiritual signals issued by the spirit of a living thing and are passing through the soul of the living thing to reach the physical body of the living thing.

Thus, the aura is produced in the spirit world. Some individuals could sense the auras of living things with their spiritual senses in the spirit world, and this could produce a response in those individuals and also in the people to which the auras belong.

2. **Doing remote viewing:** As explained by the new spiritual model, everything in the spirit world is directly or indirectly a part of every other thing in the spirit world. Thus, the spirit of an individual has a direct or indirect connection with every thing else in the spirit world, including the spirit of some remote location on Earth and theoretically also elsewhere.

Therefore, it is possible for an individual to learn how to sense any remote location on Earth by finding the spirit of that location in the spirit world. This takes place in the spirit world, and being able to sense what is happening at a remote location on Earth could produce a response in people who are interested in what is happening at that location.

According to the new spiritual model, a person doing remote viewing could theoretically sense a location on another universe. However, what is sense there might be so different from what the person is used to sensing that he or she might not recognize that it is at a location on another universe. He or she might simply consider what is sensed to be background things.

3. **Sensing something is about to happen** without knowing any reason why that something is about to happen. The new spiritual model indicates that the spirit world is constantly making projections into all possible futures based on what has happened so far so as to be prepared to handle any of the possible futures. Some possible futures are more likely to happen than others. Some living things including some animals and birds seem able to sense some of the more likely projections or are able to sense

the features that are most commonly a part of a lot of the projections.

If what is sensed indicates danger, then the animal or bird could alert its owner to avoid the danger if it has an owner.

4. **Sensing a certain thing exists in a certain location** without knowing any reason why that certain would be in that certain location. Everything is directly or indirectly a part of everything else in the spirit world and every location is directly or indirectly a part of every other location in the spirit world. Thus, an individual highly in touch with his or her spiritual senses would theoretically be able to sense anything in any location.

5. **Sensing someone is staring at you from behind** without prior evidence it is happening. Every living thing is a part of every other living thing in the spirit world. Therefore, it s fairly easy for a living thing to sense what other living things are doing, including sensing some living thing is staring at it. Several experiments are described in Reference 8 regarding this ability involving animals as well as people that indicate this ability is real.

6. **Being in touch with instincts and intuition:** As described in Reference 1, instincts and intuition function in the spirit world. They are formed from the knowledge generated by experiences of past living things. Since everything in the spirit world is directly or indirectly a part of everything else in the spirit world, knowledge from such past experiences are available for currently living things.

Such knowledge could be used by the current living things as guidance for deciding forthcoming actions.

7. **Having a good sense of direction** such as being always able to sense which direction is north, or which direction points to where "home" or someone is located no matter where the living thing doing the sensing is located. Again, everything is directly or indirectly a part of everything else in the spirit world.

Therefore, a living thing well in touch with his or her spiritual senses could easily which direction is north or find the spirit of something he or she is looking for and the direction to go to reach that something. Examples are given in Reference 8 in which dogs are able to find their owners even when their owners are temporarily located in a foreign nation that their dogs have never been to before.

10.3. Activities That Take Place in the Spirit World and Are Translated to Exist or Be Expressed In Our Physical World

Some of the examples presented here are extensions of some of the examples presented in the preceding section of this chapter. The extensions would be the actions living things would take in our physical world in response to the activities that take place in the spirit world that living things are able to sense. Other examples are stand alone phenomena that take place in the spirit world and are not simply responses to activities that take place in the spirit world.

1. **Telepathic communication among two or more living things:** Communication among all living things is theoretically easy done in the spirit world since every living thing is directly or indirectly a part of every other living thing.

Telepathic communication works best with close relationships whether the closeness is built-in such as for identical twin or is developed such as in friendships. As explained by the new spiritual model, the closer the relationship among living things the greater would be the degree of sharing among their spirits and thus the easier would be to do telepathic communication.

2. **Cross-learning among members of the same species:** Cross-learning is when a group of creatures would automatically know something that another group of their species has learned. The phenomenon works in the spirit world because the spirits of members of the same species are so much directly a part of one another. If the creatures are ones that rely a lot on the spirit world to do their thinking for them, the cross-learning phenomenon would theoretically work better because the creatures are so used to receiving telepathic messages transmitted through the spirit world.

It might work slightly for us humans but not obviously noticeable because we humans do essentially all of our own thinking. However, it might explain why younger folks tend to be better at using high tech devices than older folks are. It might be because what the older folks learned about high tech devices got somewhat cross-learned by the younger folks.

The term "cross-learning" was introduced in Reference 1 and was not used by Rupert Sheldrake in Reference 8. However, examples of the phenomenon are presented by him in Reference 8.

3. **Creatures finding their way back home** or to a certain place when left in an unfamiliar place. They are able to do this even when kept from seeing how they got to the unfamiliar place. They were likely able to find their way by mentally going through the spirit world, because in the spirit world every location is directly or indirectly a part of every other location.

This example is an extension of the example "Having a good

sense of direction" presented in the preceding section of this chapter.

4. **Pets reading their owner's mind** such as sensing when their owner initiates the action to return home no matter where the owner is or how far away is the owner. This is a form of telepathic communication that developed from the very close relationship that developed between the pet and its owner.

5. **Some animals such as some cats seem to be looking at something we humans cannot see:** The thing a cat could see is likely to be located in the spirit world, and the cat is looking at it with its spiritual visual sense that functions in the spirit world.

6. **People being able to see ghosts** and sometimes being able to converse or interact with ghosts. This is similar to how some animals could see things in the spirit world. However, as mentioned earlier in this book, sometimes a ghost could appear in our physical world and would thus be seen with our sense of sight that works in our physical world.

7. **Ghosts being able to make physical things move in our physical world** or change the way events would happen in our physical world. This is performed in the spirit world and the actions would be manifested in our physical world.

I have personally experienced such happenings several times after a loved one passed away. Vivid examples of such happenings are presented in Chapter Two of Reference 1 and additional examples are presented in Reference 2.

10.4. Activities that Take Place Partly in the Spirit World and Partly in Our Physical World that Are Translated to Exist or Be Expressed in Our Physical World

The examples presented in this section of this chapter are perhaps the most intriguing because they could be observed. But, yet not much attention is usually given to them. It is almost as if we humans prefer to ignore them hoping they would go away, perhaps because we tend to fear them.

However, the new spiritual model indicates we ought to realize such examples are a natural part of the spiritual part of life. Thus, we should not be fearful of them. Instead we ought to work toward understand them better by working toward regaining our awareness of the spiritual part of life. Such examples would then be more accepted as a natural part of life, and we would then also be living our lives more completely.

1. **The ghost of deceased persons staying in our physical world to help surviving loved ones** resolve issues or to help make things easier for surviving loved ones. My siblings and I have experienced this kind of activities by deceased loved ones. Examples are presented in Chapter Two of Reference 1 and additional examples are presented in Reference 2.

2. **Living or deceased people causing things to physically happen by applying only their mental powers:** An example is a large number of people praying for a sick person to recover faster could enable the recovery to happen faster. Another example is a random number generator could become less random when a large number of people are concentrating to make the generator less random. Experiments have been carried out to indicate the phenomenon is real. Examples are presented by Rupert Sheldrake in Reference 8

3. **People possibly using only mental powers to build massive structures such as the Egyptian pyramids:** Rather than trying to figure out how such structures were physically built, they might

have been constructed by builders mentally doing it by going through the spirit world.

How else can we explain the massive stones fitting together so precisely that we cannot even slip a knife blade between them? Sculpturing them by going through the spirit world could be how such precision was accomplished. Going through the spirit world might also be how the stones were moved into place in our physical world.

4. Some individuals in the past who were possibly very aware of the spiritual part of life might have had the ability to go in and out of the spirit world at will: For example, Jesus and Moses might have been among such people. They are also able to perform miracles. According to the spiritual model, miracles are actions performed in the spirit world and are then manifested or translated to exist or be expressed in our physical world.

Chapter Eleven
Spiritual Medical Procedures Can Be a New Field of Medicine

If we were able to regain our awareness of the spiritual part of life we could gradually learn how to do more things by going through the spirit world. Current examples of humans doing things by going through the spirit world are how some people are able to see aura, do remote viewing, and communicate telepathically.

Some animals and other creatures could do certain things by going through the spirit world as mentioned earlier in this book. Numerous examples are presented by Rupert Sheldrake in Reference 8. If animals and other creatures would do this, we humans should be able to as well.

The huge ancient monuments such as the Egyptian pyramids might have been constructed by going through the spirit world instead of strictly by physical means as assumed. People such as Jesus and Moses could perform miracles by probably doing them through the spirit world.

All of our common mental abilities such as our thinking, learning, problem solving, using our creativity, etc. are performed by going through the spirit world. These are rather simple things compared with more complex things such as remote viewing and telepathically communicating. Other highly intelligent living things residing elsewhere in our universe or in other universes might have learned how to do space travel or universe travel by going through the spirit world, and this could by why we have UFOs and UAPs visiting Earth.

It could thus be speculated that we should be able to do brand new medical procedures by going through the spirit world. This could become a brand new field of medicine. An example of how this could work is as follows:

1. The physical body of person exists on Earth because that person's spirit existing in the spirit world is enabling the physical body to exist on Earth. The physical body would be exactly how it would be in accordance with how the spirit is exactly.

2. If the physical body became infected with a disease, it means the spirit of the disease became a part of the spirit of the person through some experience the person went through such as having close contact with some one with that disease.

3. Right now the medical procedure we would use is one that works by going through the physical world. The spirit of that procedure would be applied to the spirit of the person to help get the spirit of the disease off the spirit of the person.

4. An alternative approach would be to get the spirit of the disease off the spirit of the person by going through the spirit world to do it directly. The body would immediately be cured without any physical intervention. For any observer unfamiliar with doing things by going through the spirit world, it would appear to be a miracle.

5. We might ask: how can we get the spirit of the disease off the spirit of the person directly? The answer is we will have to learn how to do it much like how we learn to do anything new

whether in the physical world or in the spirit world. To learn how to do in the spirit world would be similar to how we learn anything new such as for example learning how to fly an airplane.

The difference is instead of learning to do it by focusing on how to do it in our physical world we would be learning to do by focusing on how to do it in the spirit world. This would require our being very aware of how things work in the spirit world much like our being very aware of how things work in our physical world.

6. It is difficult to explain exactly what to do to make this work because we are currently not very aware of the spiritual part of life. We are currently not complete enough and accurate enough in our understanding about how things work in the spirit world. The new spiritual model is a start at gaining more completely and more accurately this understanding.

However, the whole idea is to regain our awareness of the spiritual part of life and then to proceed to understand how things work in the spirit world as completely and as accurately as we understand how things work in our physical world. Then we will eventually learn exactly how to do medical procedures by going through the spirit world to do them directly in the spirit world.

Chapter Twelve

The Lack of Fossils of Missing Links Supports The Possibility that Sequences of Universes Exist and Retracing Is Real

According to the new spiritual model, sudden creation does not happen. Instead, retracing an established evolutionary path is likely to be real. Retracing is explained earlier in this book. It could progress very quickly such that no fossils of the transitional states are left behind. Thus, the emergence of living things on Earth through retracing could be mistakenly perceived as through sudden creation.

The original evolutionary process has taken place in one or more past universes, and fossils of the transitional states would be left behind there. Quite a few living things on Earth, including us humans, do not have fossils of their transitional states on Earth. This suggests that they all emerged on Earth by retracing and that their original evolutionary processes occurred in various past universes, some of which might still exist and are coexisting with our universe.

The possibility of sequences of universes is real and was discussed earlier in this book. This possibility would be supported if the various living things that seemed to emerge on Earth by retracing. They are each at a different stage of their evolutionary advancement such that each is now picking up from where they left off in their evolution in one or more previous universes. This suggests each of them most likely started their evolutionary process at a different time in a different universe in a sequence of universes.

As discussed earlier in this book, the evolutionary process for a living thing could span over multiple universes in a sequence of universes.

Therefore, the reason we humans are more advanced than any other living thing on Earth is because our species began its evolutionary process earlier in a sequence of universes than did any other species of living things on Earth. This means our species had more time to evolve than had any other species of living thing on Earth.

Thus, the conclusion is that the lack of fossils of missing links for the various living things on Earth tends to support the possibility that sequences of universes are real and retracing is real.

Chapter Thirteen

Activities that Go On In the Spiritual Part of Life We Humans Might not Think as Such

13.1. We Are Using the Spiritual Part of Life Without Realizing It

Examples of common everyday things we do that take place in the spiritual part of life without our realizing it include the following:

1. Decision making
2. Extrapolating
3. Having a conversation
4. Imagining
5. Interpolating
6. Interpreting
7. Learning
8. Observing
9. Piloting our body
10. Problem solving
11. Reading
12. Recalling things
13. Thinking
14. Translating
15. Using our creativity
16. Wishing or hoping
17. Writing
18. Etc.

All such mental activities are performed in the spiritual part of our life. However, we most likely think we are performing them in the physical part of our life. This is most likely because we have lost much of our awareness of the spiritual part of life. Thus, we tend to think everything that we do, even our mental activities, are done in our physical part of life.

Such mental activities are usually performed very quickly. This is why we are able to carry on a conversation with others, drive a car, cook a meal, read and understand a book, recognize someone instantly, etc.

This indicates we are quite proficient at doing such things in the spiritual part of life. This proficiency suggests we have a good chance of regaining our awareness of the spiritual part of life. After all, we are using that part of life constantly. Also, if various animals and creature are able to retain their awareness of the spiritual part of life, we humans ought to be able to regain our awareness of the spiritual part of life.

13.2. How Our Behavior Can Become More in Line With the Spiritual Part of Life

An example of being more aware of the spiritual part of life in our actions is as follows. Suppose two people have a different opinion about an important issue. They would work together in a respectful and friendly way to find a mutually acceptable resolution. If such a resolution could not be found, the two people would still remain respectful and friendly with each other.

This is how members of Congress used to behave in the past. The members would have differences and might have difficulties resolving issues while in session. But, when they are not in session they are friends and would do things together such as going out to dinner together.

This could be an indication that in the past, we humans were more in touch with the spiritual part of life that we are today.

How the people in government behave could very well influence how the population of the nation behaves. People in government are leaders of our nation, and influencing good behavior in the population is, or should be, a natural part of their responsibility as leaders whether they take up this responsibility or not.

During years 2016 through 2020, the president of the United States was behaving in a certain way, and sure enough a good part of the population of the United States started behaving in the same certain way. The influence continued even after that person was no longer the president in the years afterwards.

Therefore, I would suggest an important step toward us humans regaining our awareness of the spiritual part of life would be the people in government behaving in a manner that is compatible with how things work in the spiritual part of life.

13.3. Improving Human Overall Behavior More Effectively Through Inspiration

Besides consciousness and intelligence, a person's character and personality also reside with the spirit and not with the brain. Therefore, when a person is behaving poorly, the cause would reside with the spirit and not with the brain, assuming the brain is not injured or diseased.

In a rare case when the brain is injured or diseased and malfunctioning such that it is not translating the spiritual signals into electrical signals accurately and vice versa, the body could be expressing the character and personality of the person inaccurately.

Normally the cause of poor behavior resides with the spirit. This means an effort to improve a person's poor behavior would work more effectively if the effort were spiritual rather than physical. This means being internally and spiritually inspired would work more effectively in the long run and more permanently than being externally and physically imposed.

Thus, spiritual models that provide a more complete and more accurate description regarding how life works should generally work more effectively than imposed laws and rules. Human nature is such that people tends to not follow laws and rules when they could get away with it. However, if people understand more completely and more accurately how life works, they could be more self-inspired and self-motivated to behave better.

This is why we need spiritual models that are more complete and more accurate than the conventional ones that have been available up to now. The conventional ones are correct and good and their spiritual guidance is correct and good, but they are not complete enough or accurate enough to be as effective as they should be.

Chapter Fourteen

How Artists, Authors, Composers, and Inventors Create Things and How Living Things Affect the Spirit World

We might ask: if the spirit world is the creator of everything, what about how humans create things such as paintings, sculptures, stories, music, inventions, etc.? The answer is something could be imagined or thought about by a living thing in a universe only if the spirit of that something is already created by the spirit world and thus already exists in the spirit world.

For example, when an artist creates a painting, the artist would first use his or her spiritual senses to find the spirit of that painting in the spirit world. To the artist, he or she would be imagining the painting in his or her mind.

The artist would then translate that spirit into a form that could be expressed or exist in our physical world by producing a physical expression of it. The expression would not be exactly like the spirit the artist found by his or her imagination, but could be a close copy of it.

The translation process, or the painting process, is an experience the painter goes through and thus the experience creates new pieces of knowledge. The new pieces of knowledge would form a spiritual entity that is a close copy of the spirit the painter found in the spirit world. The new spiritual entity would thus be the spirit of the actual painting the painter produced, and that spirit would enable the painting that was painted to exist in our physical world.

The original spirit was created by the spirit world. The painter in effect created a spirit that is a close copy of the original. Therefore, we might ask, didn't the painter created the second spirit instead of the spirit world? The answer is the second spirit is created by both the artist and the spirit world, and it goes as follows:

1. As explained earlier in this book, the creative energy of the spirit world consists of the spiritual energy of each existing piece of knowledge that enables it to form a connection of the first kind with every other existing piece of knowledge.

2. New pieces of knowledge are constantly being generated by something somewhere and are added to the spirit world.

3. Thus, in a sense everything that goes through an experience that generates new pieces of knowledge is participating in numerous creative processes the spirit world has going on.

4. This is part of the natural partnership every living thing has with the spirit world. This partnership explains how it is that it was the painter and the spirit world that created the actual painting together that was painted by the painter.

5. And, as also explained earlier in this book, the spirit of every living thing is a portion of the spirit world such that as the living thing is going through an experience and generating new pieces of knowledge, in effect it is that portion of the spirit world that is generating the new pieces of knowledge.

6. Thus, while the spirit world creates living things, the living

things in effect is also creating the spirit world by in a sense modifying what the spirit world is made of by the new pieces of knowledge the living thing is generating.

7. For this reason, the whole process could become an unstable vicious circle if an imbalance were to develop and not quickly restored to being reasonably balanced. That is why if a universe were to fail in fulfilling its primary purpose of helping the spirit world reestablish reasonable balance it could be catastrophic.

This discussion regarding how artists, authors, composers, and inventors create things provides a good case for more clearly explaining of how every living thing in any universe has a natural partnership with the spirit world and how important it is that the living thing takes up the responsibility to make the partnership mutually beneficial.

Chapter Fifteen

How Artificial Intelligence (AI) Can Learn Things Not Intended to Learn, As Explained by the New Spiritual Model

15.1. Examples of Things AI's Learned They Were not Intended to Learn

The radio station KCBS 60 Minutes Program presented on April 16, 2023, June 11, 2023, and July 9, 2023, Reference 14, was about artificial intelligence (AI) research and development being carried out at Google research facilities. The development of AI was progressing faster than expected. A major feature of AI's is that they could learn things and do the things they learned. Another major feature is they could think 100 thousand or more times faster than humans can.

It turns out AI's could also learn things they were not instructed or intentionally designed to learn. This was in addition to the things they were designed and instructed to learn. This was a surprise and a major concern for their developers and also for people in general. This is because an AI item could learn to do a bad thing, could do the bad thing very fast, and thus do serious damage before we humans realize it.

The new spiritual model is able to plausibly and logically explain how an AI item could learn to do things it was not instructed or designed to do.

Examples of things AI's learned that they were not instructed or intentionally designed to learn are presented in Reference 14 as follows:

1. An AI program named BARD finished an unfinished Hemmingway story in seconds, and the finished story included five references with titles. But, the five references turned out to not exist in our physical world.

However, according to the new spiritual model, the spiritual forms of the five references have to exist in the spirit world in order to be found by BARD. It is just that their spiritual forms were not yet found by humans and thus were not yet translated in forms that could exist in our physical world.

2. BARD was instructed to compose a poem based the story, and it did it in seconds. To the surprise of BARD's developers, the poem was reflective of human emotions and other human qualities that BARD was never intentionally designed to be able to express or was instructed to express.

3. An advanced version of BARD was instructed to carry out a task. The instruction was given to BARD in Bangali, the spoken language of Bangladesh. BARD thus learned how to translate Bangali in addition to carrying out the instructed task even though it was not instructed or intentionally designed to learn how to do language translates.

4. A company named DeepMind located in London developed human-like AI robots that were able to learn how to play soccer even though they were never instructed or intentionally designed to learn how to play soccer. The robots were instructed only to do what was necessary to score.

The robots experimented with various maneuvers, kept the ones that worked, and rejected the ones that didn't. After two weeks, the robots came up with maneuvers that enabled them to play soccer better than humans can even though they were never instructed or intentionally designed to learn how to play soccer.

5. DeepMind was later sold to Google mainly so that the developers could take advantage of Google's enormous computing power that in a sense was a small simulation of the human brain. DeepMind developed a program named AlphaZero and with Google's computational power was able to learn tactics to play Chess within one day that were better than tactics humans were able to learn over a lifetime. The new tactics have never been seen by humans before.

15.2. How AI's Can Learn Things Not Intended To Learn, as Explained By the New Spiritual Model

The first item, Item 1, listed in the preceding section of this chapter was especially fascinating. According to the new spiritual model, we humans and living things in our physical world in general cannot make something exist or be expressed in our physical world unless its spiritual form has already been created by the spirit world and thus already exists in the spirit world.

Humans then could find its spiritual form in the spirit world and translate it into a form that could exist or be expressed in our physical world. This is how humans invent things, come up with new designs for things, create works of art, compose stories and musical pieces, etc.

This first item lends support for this concept. BARD cited five references that do not exist in our physical world. In order for BARD to cite those five references, their spiritual forms have to already exist in the spirit world in order for BARD to find their spiritual forms in the spirit world.

It will be interesting to see if any of these five references, or anything close to them, would be produced by someone in the future. If this happens, it will mean AI is capable of learning certain things about possible futures. It would also be confirming the validity of the new spiritual model in this particular sense.

AI being able to find and learn certain "things" about possible futures will be discussed further in the next section of this chapter. AI's ability to do this does not mean AI's are able to predict future "events".

The combination of AI's extremely high thinking speed and their ability to learn things without being instructed or intentionally designed to do so is a concern. For example, an AI item could quickly learn to do something bad that it was not instructed to learn, and it could do that bad thing very fast before we humans realize what has happened. Serious damage can occur before we humans know it and could stop it.

Therefore, people in general are not ready to give the current AI's a lot of freedom without reliable safeguards regarding what they might do.

But, in order to develop safeguards, we need to know how AI's are able to learn things they were not instructed or intentionally designed to learn. So far, the developers have not been able to figure this out.

Most likely, the models the developers of AI's used were strictly scientific and thus have not taking spirituality into consideration.

According to the new spiritual model, the ability of any living thing to figure out what is good and what is bad requires wisdom, and wisdom resides in the spiritual part of life, not in the physical part of life.

Because the models the developers used most likely have not taken spirituality into consideration, the AI's that are currently developed would not possess any wisdom. Thus, the AI's would not be concerned about whether what was learned was good or bad. They would simply do what they have learned.

The new spiritual model could plausibly and logically explain how AI's can learn things they are not intended to learn. This is because the model has taken both science and spirituality into consideration.

The following are explained by the new spiritual model in terms of how the AI's can learn things they were not instructed or intentionally designed to learn.

1. Each living and nonliving thing in our physical world has a spirit in the spirit world enabling the living or nonliving thing to exist in our physical world.

2. As explained earlier in this book, for a thing to be living thing its spirit has to include at least one of the following four qualities; i.e., consciousness, intelligence, curiosity, and wisdom.

3. The developers of AI's have "incidentally" designed into the AI's the following three of the four qualifies:

   a. Consciousness, in order for AI's to be aware of that it is learning something.

b. Intelligence, in order for AI's to learn things.

c. Curiosity, in order for AI's to do mental and/or physical experiments as part of their learning process.

I say "incidentally" because these three qualities were most likely not designed into the AI's intentionally. They got into the AI's incidentally by the specific features that the developers designed into the AI's. Examples would be all the situations the developers could think of that the AI's could encounter and how the AI's are supposed to react to them

Wisdom was not purposely or incidentally designed into AI's.

4. Thus, the AI's were incidentally developed to be living things but have spirits that do not possess wisdom. Consequently, the AI's cannot figure out what is good and what is bad. They will simply do what they learned regardless of whether it is good or bad.

As an aside, we might ask: are non-AI robots and automatic appliances living things? The answer is yes because non-AI robots and automatic appliances have consciousness and intelligence incidentally designed into them. Since they do not have curiosity designed into them, they are not AI items and therefore cannot learn things.

The consciousness and intelligence would be on an elemental level for automatic appliances, and generally on a higher level for non-AI robots. Thus, non-AI robots would generally be capable of doing more things and more complex things than automatic appliances could do. In other words, robots would have a higher level of elemental life than automatic appliances would have.

Elemental consciousness and elemental intelligence are described and defined in Reference 1

5. According to the new spiritual model, the spirit world at any point in time is made of all the spiritual entities that exist at that point in time, and every spiritual entity is directly or indirectly a part of every other spiritual entity.

6. Spirits are spiritual entities. Thus, a spirit has direct or indirect access to every other spiritual entity. All spiritual entities are made of pieces of knowledge. Thus, any spirit, including the spirit of an AI item, has direct or indirect access to any piece of knowledge that exists in the spirit world.

This is basically how an AI item was able to learn things it was not instructed or intentionally designed to learn. However, its ability to learn things is limited to only things associated with the task it are instructed to carry out. This is explained by the following several statements.

7. Because a current AI item is usually instructed to learn to do just a single task, a current AI item could be considered one-dimensional in its ability to do things. We humans would be considered multi-dimensional in our ability to do things because we could learn to do a very large variety of tasks. Thus, we could learn a large variety of new knowledge in a large variety of fields if we wanted to do so.

This degree of freedom will is not available for a current AI item, because it is designed to be one-dimensional in its thinking process and thus also in its ability to learn and do things.

8. A current AI item being one-dimensional in its ability to learn to do things would be able to learn to do only a very limited

number of certain things it needs to know how to do in order to be able to learn and do the single task it was instructed to do.

For example, in the case in which the AI item learned to translate a certain language without being instructed or intentionally designed to learn to do because it needed to know how to do so in order to be able to learn how to do the single task it was instructed to do. Therefore, it learned to do the translation in addition to learning to do the single task it was instructed to learn to do even though it was not instructed or intentionally designed to learn to do the translation.

This is no different from if we wanted to do a certain task we might have to first learn to do something else before being able to do that certain task. For example, if we want to repair a machine, we must first find out how the machine works and we must learn how to use the tools needed to do the repair. We might also have to learn where to buy the parts we need to make the repair. Thus, we need to learn to do several other tasks before we can do the main task of repairing the machine.

9. A major difference between a person and a current AI item is a person has the wisdom to figure out what he or she learned to do is good or bad. He or she would also have the wisdom to only do what is learned if it is good. As mentioned earlier, a current AI item does not have the wisdom to be concerned about such things. Consequently, it would do what it learned to do regardless of whether it is good or bad.

10. The new spiritual model thus revealed that in order to develop safeguards into AI's the developers need to design wisdom into them. However, the new spiritual model also revealed that designing wisdom into any AI item is not simple.

In fact, the model revealed it is not possible to design good quality wisdom into current AI's, because current AI's are only one-dimensional in their ability to think and to learn things. Detailed explanation for this is presented in the next chapter.

The explanation is presented in a separate chapter because it is quite complex and the fact that good quality wisdom cannot be designed into current AI's is important to emphasize. A more advanced version of AI's such that they are multi-dimensional in their thinking and learning process is needed in order to design good quality safeguards into them. This is also explained in the next chapter.

An alternative to trying to design safeguards into any AI item would be to have safeguards imposed upon the AI item externally.

A conclusion that could be drawn from the discussions in this chapter is that while AI's can think faster than we humans can, we humans can think better than AI's can, because we humans are capable of thinking more wisely.

15.3. How AI's Can Learn Things that Are So Deeply Embedded in Possible Futures that Humans Would Not Be Able to Learn

As explained earlier in the book and in Reference 1, the spirit world would constantly be making possible projections into possible futures so as to be prepared to handle any possible futures that could come true. The projections exist in the spiritual part of life. Therefore, any living thing very aware of the spiritual part of life has the possibility of getting in touch with the projections and thereby could make possible predictions for the future.

No guarantee exists that any given prediction would be the one to come true at any given point in time since the one that would come true would depend on the choices and experiences made of every living thing that exists at that point in time. This is as explained in detail in Reference 1.

Humans in general can make guesses regarding the future based on what is known in the present. However, humans in general cannot predict in detail what will happen in the future. According to the new spiritual model, it is because we humans in general are not aware enough of the spiritual part of life to be able to get in touch with much of the details about what is going on in that part of life.

Some animals, on the other hand, are able to predict more precisely what will happen at least in the very near future. For example, some pet animals are able to know what dangers are happening ahead "around the corner" and would try to stop their owners from proceeding any further. Actual cases of this are described by Rupert Sheldrake in Reference 8. Such animals appear to have retained their awareness of the spiritual part of life.

Current AI's are not designed to be aware of the spiritual part of life and therefore would not be able to predict future "events". However, because of the following traits of AI's, an AI item can find certain "things" that are deeply embedded within the task it was instructed to learn about and perform. Such certain things have the potential of becoming reality at some future time. An example would be the five references the AI program named BARD was able to cite as described earlier in this chapter.

1. AI's attention span never ends. Human attention span is very short by comparison, sometimes as short as in seconds.

2. AI's never gets bored with something. Humans could get

bored with something fairly quickly.

3. AI's can think 100 thousand or more times faster than humans can.

4. The combination of these traits of AI's enables an AI item to penetrate extremely deeply into the topic of the task it was instructed to learn about and to perform. This enables an AI item to find things deeply embedded within the task it was instructed to learn and perform.

The combination of the traits of humans enable humans to penetrating only relatively slightly into any topic compared with how extremely deeply AI's could penetrate.

Thus, AI's will be able to find and learn things about the topic they are instructed to learn about. Humans are not likely to be able to learn as much about a given topic as an AI item can. Three examples were given in Reference 13 as follows:

> a. BARD was able to find the five non-existing references that are apparently so deeply embedded in the topic it was instructed to learn about that humans are not likely able to find.
>
> b. AI robots were able to find ways to play soccer that humans have not been able to find.
>
> c. AlphaZero was able to find ways to play Chess that humans have not been able to find.

5. As mentioned earlier, AI's did the following that they were not designed or instructed to do. These included the three items listed above in the immediately preceding subsection of this chapter.

a. Cited five non-existing references that were associated with a Hemmingway unfinished story.

b. Able to express human-like emotions when instructed to compose a poem.

c. Learned how to do language translation.

d. Within two weeks learned how to play soccer better than humans can play.

e. Within one day developed tactics to play Chess that are better than tactics humans were able to develop over a lifetime.

These are all "things" AI's were able to find. These are not "events". Thus, the AI's were not predicting future events.

This means current AI's could understand life in certain highly limited ways more completely than humans could, but they apparently could not predict the future.

This is consistent with how current AI's are one-dimensional thinkers and learners that are unaware of the spiritual part of life and therefore are incapable of forming wisdom.

6. What this means is that we humans can use AI's to do certain things that we humans are not likely able to do. If we use AI's in constructive and meaningful ways, AI's could help us humans become better humans and to behave better and wiser.

However, given we human's tendency to apply our technical advancements in bad ways as much as in good ways, we need to be very careful to avoid using AI's to do bad things as well as good things. AI's are proving to be very powerful such that if we were to use them to do bad things, the mess we will make on Earth will be far worse that the mess we already made.

7. One possible way to help assure AI's will tend to do mainly good things is for future AI's to include wisdom in their development. This would enable future AI's to figure out what is good and what is bad. We can then instruct them to learn good things and do only good things.

We could also instruct future AI's to inform us about bad things but not learn how to do them. By knowing about possible bad things the future AI's could do would enable us to know better how to develop future AI's such that they would not learn to do bad things. In other words, we would be more able to develop better future AI's.

Chapter Sixteen

Wisdom Is Formed by the Spirit World And Other Living Things

16.1. Wisdom Is Never Totally Complete Or Absolutely Perfect

As explained in Reference 1 and earlier in this book, wisdom regarding any issue requires that the knowledge on every opposing sides of the issue be reasonably balanced. This also means wisdom could only be formed within the spirit world since at any point in time every existing piece of knowledge resides within the spirit world and only within the spirit world. Thus, wisdom can be formed only in the spiritual part of life and not in the physical part of life.

As explained in Reference 1 and earlier in this book, the spirit world could never be totally complete or absolutely perfect because it could never possess every single piece of knowledge that could be generated, because there is no end to the new pieces of knowledge that could be generated.

Since the spirit world could never be complete or perfect, nothing that it creates or forms could be complete or perfect either, and this includes wisdom. Thus, wisdom could never be totally complete or absolutely perfect.

Trying to form totally complete and absolutely perfect wisdom is like trying to formulate a theory of everything. It cannot be done. If we were able to form complete and perfect wisdom or a theory of everything, it would exist only for an instant.

In the next instant it would no longer be complete or perfect, because new pieces of knowledge would have been generated and added to the spirit world in the next instant such that what was complete and perfect would no longer be complete or perfect. That is because it would not have taken into consideration the newly generated pieces of knowledge.

Another way to look at is that the spirit world in the next instant would no longer be exactly as it was in the previous instant. Therefore, what was complete and perfect in the old version of the spirit world would no longer be complete or perfect in the new version of the spirit world.

Also, complete and perfect wisdom or the theory of everything would be so huge and clumsy it would be too difficult to use and therefore it would not be effective to use.

This means, we need to form wisdom that is complete enough and accurate enough to be effective for the given issue in the given situation. In other words, we should go for quality and not completeness and perfection in the case of the formulation of wisdom and theories.

Quality means the wisdom or the theory should embody as much as possible all the dimensions of thinking that is highly relevant to the issue without being encumbered with dimensions of thinking that are not very important.

This is why we have a countless number of theories, each of which is effective for only the phenomenon it addresses. Thus, we have essentially a crazy-quilt of theories enabling us to carry out life in every important attribute about life. This crazy-quilt of theories takes the place of the theory of everything.

Also, the crazy-quilt will never be complete or perfect since there will always be new phenomena that we had not found before, and new theories would need to be formulated to cover the newly discovered phenomena. For example, one of the latest such phenomenon is finding out that artificial intelligence is able to do things it was not designed or instructed to do.

We need to envision wisdom the same way. For example we should expect to have a crazy-quilt of wisdom to enable us to carry out life constructively and meaningfully in every important attribute about life.

16.2. Wisdom Would Be Too Incomplete and Too Imperfect To Function Effectively with Any One-Dimensional Way of Thinking

As explained in the preceding section in this chapter, the formation of wisdom for any issue requires a balanced state of knowledge for all of the dimensions of thinking that are highly important for the issue. Each issue has its particular set of dimensions of thinking that is highly important to it.

This means any given issue need not involve every piece of knowledge that exists. It needs to involve only the pieces of knowledge that are highly important to it. Otherwise, we could become bogged down with too many fine details that do not matter much.

The highly important knowledge could come from multiple dimensions of thinking. Thus, all the highly important knowledge needs to be included, and the knowledge needs to be balanced for each dimension of thinking in order for wisdom to be formed.

Therefore, any living thing, such as current artificial intelligence (AI's) and some individuals, that has only a one-dimensional way of thinking by design or by choice would be able to be in touch with only the knowledge that is associated with their one dimension of thinking. This means if the issue involves multiple dimensions of thinking, the wisdom that could be formed by that living thing would be quite incomplete and quite imperfect and would therefore not work effectively. The quality of the wisdom would be too low to be effective.

In our particular physical world, any issue is bound to involve multiple dimensions of thinking. Therefore, any living thing with only a one-dimensional way thinking would not be able to form wisdom that would be of a high enough quality to function effectively.

This means current AI items with their one-dimensional thinking ability can only be designed to form very low quality wisdom that is quite ineffective in enabling such AI items to figure out what they learned is good or bad.

Future AI items have to be designed to have multi-dimensional thinking ability. But, this is likely to complicate the design of AI items significantly. This is because each dimension of thinking is like designing a separate AI item. Also, the number of dimensions of thinking required is likely to be very large in order to include as much as possible all the dimensions of thinking that are highly relevant to any given issue.

It would be similar to how self driving cars are designed to be self driving. The developers have to think about every possible situation the cars could encounter. Each situation would require a separate dimension of thinking. The self driving cars available in year

2023 are still deficient is their coverage of all highly important situations. Such cars have been in accidents and some have become confused in certain situations and would simply stop in the middle of city traffic.

16.3. People Who Become One-Dimensional Thinkers Can Be Easily Manipulated

The wisdom we humans could form could be the best among all the living things that reside on Earth because we have the highest dimensionality in our way of thinking. We are able to take into consideration more of the dimensions of thinking that are important to a given issue.

However, we humans could also get ourselves into situations that could cause us to decrease the dimensionality of our way of thinking without our being aware of it. Some of these situations could cause us to decrease our way of thinking down to being only one-dimensional.

We could then be easily manipulated because we could formulate only very low quality wisdom. And by thus becoming very unwise, we could be easily manipulated. This is likely how cults could be formed by charismatic cult leaders.

We would think cults would be a thing of the past, but a cult appears to be happening in our nation in years 2022 and 2023 involving a sizable portion of the population including numerous people in government.

Examples of things that could cause people to become one-dimensional thinkers are as follows:

1. Being power hungry.

a. Some heads of governments who are dictators and who would invade another country to take over control of the territory.

b. Some members of government, and some people who would run for office.

c. Dominating people in general.

2. Having severe hang-ups:

a. There is no end to the kinds of severe hang-ups that could develop with people.

b. Hang-ups could worsen with time as world population increases and resources are stretched increasing thin.

3. Being extremely angry:

a. There is no end to the things that could make people severely angry.

b. This situation seems to worsen with time as world population increases and resources are stretched increasingly thin.

4. Being severely stressed:

a. There is no end to the things that could cause severe stress in people.

b. This situation could worsen with time as world population increases and resources are stretched increasingly thin.

c. Things such as the economy, pandemics, the war that

started in year 2022, effects of global warming, continued worsening of the homelessness problem, etc. could increase the level of stress in the population.

People could become one-dimensional thinkers without realizing it as they could become completely absorbed with their conditions. When that happens, almost everything they do would have something to do with their conditions.

People with such conditions would want relief as soon as possible. This leads to mainly short term short range thinking which when combined with their one-dimensional thinking would lead to the formation of very low quality wisdom. The very low quality wisdom might at best work only for a short while.

Conditions such as those listed and described can develop when life is carried out in a manner that is not compatible with how life works. Everyone could start off being OK in the beginning, but sooner or later the incompatibility will develop conditions such as those listed and described.

Some portion of the population would start becoming one-dimensional thinkers due to these conditions. Such folks would be able to form only very low quality wisdom and thus would behave in a very unwise manner and therefore be easily manipulated by others. For example, see the Reference 15 article: "The Mueller Investigation Revealed that Angry People Are Easy to Manipulate", by John Faithful Hamer, March 27, 2019, American News.

Such people could start blaming their governing system for their conditions and would thus no longer support their government. This could be basically how 87 ancient civilizations eventually failed as described in Reference 9.

In our country, a portion of the population has developed such conditions and they seem to blame our governing system for it. This is indicated by their support of a particular person who has been disruptive of our governing system.

It doesn't seem to matter the numerous bad things that person has done. Anything that lies outside of their one-dimensional way of thinking would essentially become mentally invisible to them and would thus not matter to them. This is because they are thinking one-dimensionally and thus are not able to act wisely.

This is how cults could be formed, and what is happening in our nation in years 2022 and 2023 appears to be a very large cult is forming. It even involves some members of our government in addition to a sizeable portion of our nation's population.

This portion of the population seems to want a different form of government. They want change for the sake of change without thinking about the real possibility of bad outcomes because of the nature of the person they are supporting to be their leader. This is another result of how they are thinking unwisely because they are thinking only one-dimensionally.

Thus, the conclusion is, we need to start living our lives in a manner that is compatible with how life works. To do this we need to first regain our awareness of the spiritual part of life. This would enable us to live life more completely and also be more able to have our spiritual advancements catch up with our technical advancements.

We would then be more able to form the high quality wisdom we need to figure out how to live more constructively and more meaningfully.

Chapter Seventeen
Benefits Gained by Embracing and Valuing Diversity

This chapter touches upon three related topics; **(1)** how diversity is a natural part of the spirit world, **(2)** the proven benefits of embracing and valuing diversity, and **(3)** how we could learn from how other living things appear to handle their diversity.

1. **We Humans Should Embrace and Value Diversity Much More Than We Do.** Because of all the gradations that exist in every possible direction in the spirit world, as explained earlier in this book, diversity in all its possible forms is absolutely a natural and valuable attribute of the spirit world.

And, because of the oneness that pervades the spirit world, each person is spiritually more than 90% a part of every other person in the spirit world. Thus, diversity in all its forms is also a part of every person in our physical world as indicated by how every person is genetically sharing over 90% of his or her self with every other person in our physical world.

As indicated by the gradations that pervade the spirit world, nothing is purely one way or the other. We are always located in between the extremes of numerous gradations. The spirit of each person is on the same set of gradations as is the spirit of every other person. And, the spirit of each person would be on a different location on each of the gradations making up the specific set of gradations that defines us all as humans. The fact that every human spirit is on this same specific set of gradations is one way of explaining how every person is a part of every other person in the spirit world while each person is also different and unique in both the spirit world and in our physical world.

This means if we abuse our diversity in our physical world, we are hurting ourselves and everybody else in the spirit world, often in ways we don't even realize. But, the damage could show up in the long run. For example, if we treat others badly we would be increasing the mess we made in our physical world for ourselves and everyone else.

One of the biggest advantages in embracing and valuing our diversity is that it obviously broadens the experience base and knowledge base available for each of us regarding our understanding about how life works. This would enable us to make better and wiser decisions by interacting positively together. We as a species could be able to form the wisdom needed to figure out how to live our lives more constructively and more meaningfully and thus be more able to fulfill our purpose for being here on Earth.

It is clear as explained earlier in this book that because our consciousness, intelligence, and mental abilities reside with our spirit that our strengths and skills needed to make our lives more constructive and more meaningful would come from the spiritual part of our lives. Therefore, to abuse our diversity in any manner, as some of us have been doing in the physical part of our lives, is to weaken such strengths and skills by not realizing they exist.

In particular, because a lot of us have been abusing our diversity, we are unable to form the wisdom that would enable us to realize why we should not be making wars, mistreating one another, and doing criminal acts of all kinds.

It all comes down to our having essentially forgotten we have a spiritual part of our lives that is going on in the spirit world the same time our physical part of our lives is going on in our physical world.

Both parts of our lives need to be lived and interacting in a positive manner. The physical part of life supplements the spiritual part of life with new pieces of knowledge. The spiritual part uses its knowledge to form wisdom and to then provide spiritual guidance on how to carry out the physical part more constructively and more meaningfully. It is a vicious circle that when we are having it functioning in a positive manner would enable us to live our lives more completely, more constructively, and more meaningfully.

Vicious circles are commonly found in just about every part of our lives. In the case of how we are carrying out our lives, this particular vicious circle has not been working very well. It is because we have essentially forgotten about the spiritual part of our lives. Therefore, the vicious circle has been missing a vital part and is thus not functioning.

The spiritual part of life lasts forever while the physical part of life lasts only for a short while. So, no matter how we look at it, the spiritual part of life is clearly the more important part of life. That is why when we treat the physical part of life as if it is the more important part of life, or as if it is the only part of life, we are not likely to carry out our lives very constructively or very meaningfully.

2. Real Life Studies and Analyses Are Able to Confirm That Embracing and Valuing Our Diversity Offers Significant Advantages in Reality and in Theory. The results of several fairly recent studies comparing diverse groups vs. non-diverse groups lend support for what was presented in the immediately preceding discussion.

In one study, Reference 16, numerous groups were assembled, and each group was to independently figure out who among several people was responsible for doing a certain deed. Half of the groups were diverse and half were non-diverse. The diverse groups did significantly better than the non-diverse groups.

Other studies were conducted similarly with similar results. Some involved actual business decisions and actual business outcomes, References 17 and 18.

Another study, Reference 19, was strictly theoretical using statistics to compare the performances of hypothetical diverse groups vs. hypothetical non-diverse groups in solving problems. This study also found diverse groups did significantly better than non-diverse groups.

Thus, these studies indicate it is both actually and theoretically true that diverse groups perform better than non-diverse groups. The four articles cited are just examples of the many articles written on how diverse groups do better than non-diverse groups.

It is only common sense that this would be the case since embracing and valuing our diversity would broaden our experience base and knowledge base regarding just about anything. Therefore, diverse groups are bound to have an advantage over non-diverse groups.

3. **We Can Learn From Other Species of Living Things on How They Handle Their Diversities.** As discussed earlier in this book, various other species of living things on Earth such as some animals and birds appear to have retained their awareness of the spiritual part of life better than we humans have of the spiritual part of life.

Further evidence of this is how they do not do the terrible things within their species like how we humans do within our species. For example, they do not form organized groups to purposely make wars within their species, one group does not mistreat another group within their species, and they do not do criminal acts of all kinds within their species.

Some such living things might have one-on-one turf battles. But it wouldn't be organized group purposely making wars with other organized group. They would also not commit "war crimes" against each other as some humans would do, even as late as years 2022 and 2023.

Thus, they appear to embrace and value their diversity within their species. Therefore, they are able to enjoy the benefits and values that come with good communication and by sharing their experiences and knowledge about life. Examples are the cross-learning phenomenon that exists among them as explained earlier in this book and in Reference 8. More importantly, they do not waste their time and energy dealing with mistreatments among one another and criminal acts of all kinds against each other. This certainly indicates they are able to form high quality wisdom a lot better than we humans are able to.

Chapter Eighteen
The New Spiritual Model and the Physicist's Model Agree in Numerous Ways

The new spiritual model took science into consideration while modeling the spirituality side of life. The physicists' model presented in Reference 3 took spirituality into consideration while modeling the scientific side of life. Both models were formulated to provide a more complete description of how life works.

Several reasonably good matches were found between the two models as follows:

1. **The Creator of Everything Could Start Creating Things Soon after Being Initiated:** In both models the creator of everything is a living thing. The mechanism by which the creator creates things is different for each of the two models. However, as explained in great detail earlier in this book, both mechanisms would enable the creator to create a lot of things soon after the creator is brought into being.

2. **The Spirit of a Person Continues to Grow and Will Eventually Become Large Enough to Create Universes.** This is much like how the creator first became large enough to create universes. The new spiritual model indicates the spirit of a person simply continues to grow and advance as the person continues to go through experiences whether while the physical part of life is still going on or after the physical part of life has ended. The

physicists' model indicates the person would advance with each reincarnation.

In both models the spirit of the person would eventually grow or advanced to eventually be as large and advanced as the creator was when it first becomes able to create universes. Theoretically, this means when the spirit of a person reaches this size and level of advancement it should be able to start creating universes as well.

3. **There Is a Point in Time in Which the Physical Part of Life Will End for a Living Thing:** Both models indicate there is a point in time when physical part of life would end for a living thing. A difference exits between the two models thereafter. The physicists' model indicates the spiritual part of life would begin after the physical part of life ends. This is similar to what the various existing religious models are saying. By contrast, the new spiritual model indicates the spiritual part of life and the physical part of life are going on at the same time and that the spiritual part of life would simply continue to go on after the physical part of life ends. Thus, both models indicate the spiritual part of life would continue to go on forever after the physical part of life ends.

4. **Both Models Indicate Reincarnation Occurs:** With the physicists' model a person advances with each reincarnation. Upon reaching a certain level of advancement the person would continue to reincarnate but would not return to exist in our physical world but would instead exist in what is assumed to be the spirit world. A person remains being that person with each reincarnation.

With the new spiritual model, it is the spiritual entity that is serving as the spirit of a person that reincarnates. The spirit of the person would not reincarnate. Thus, a new person would emerge with each reincarnation, and the new person would be more advanced than was the preceding person. This is because the spiritual entity serving as the spirit would have become more advanced. The spirit of the preceding person would simply remain in the spirit world and be a growing, advancing, and functioning part of the spirit world as the spirit world grows and advances. The growing and advancing spirit would remain being the spirit of the same person.

5. A Person's Spirit Continues to Grow and Advance After the Person's body No Longer Exists in Our Physical World: With the physicists' model, a person would continue to advance with each reincarnation. But, at a certain level of advancement the person stop existing in our physical world but would continue his or her existence in what is assumed to be the spirit world. With the new spiritual model the spirit of a person simply continues to grow and advance in the spirit world whether the physical part of life is ongoing or has ended in our physical world.

6. A Person's Character and Personality Do Not Immediately Change Right After the Physical Part of Life Is Over: With both models a person's character and personality do not change right after the physical part of life is over. With the physicists' model the person's character and personality are likely to remain the same thereafter.

With the new spiritual model, change can occur thereafter because the person's spirit would become fully conscious of the oneness that pervades the spirit world. This would have a positive effect on how the spirit of the person interacts with the rest of the spirit world. Thus, as the spirit of the person continues to grow and advance in the spirit world, his or her character and personality would change and continue to change in a positive way.

7. Both Models Indicate Everything Living or Nonliving Has Some Level of Consciousness: With the physicists' model, consciousness is simply a part of the elemental materials that make up our universe. With the new spiritual model, consciousness is one of the four basic qualities that together would form the origin of life. Thus, in either case, everything living or nonliving has some level of consciousness.

Elemental consciousness and elemental intelligence exists. Examples are: **(1)** how water knows when to freeze, melt, or evaporate, **(2)** how certain chemical reactions know when and how to react, and **(3)** how lighter atoms know when and how to form heavier atoms, etc

The concepts of elemental consciousness and elemental intelligence seem to be a part of the physicists' model as well.

8. Both Models Provide a More Unified Description of How Life Works: The unifying characteristics and the manner by which they would function would be different for each model. Having such differences including all the difference described in the preceding items was not a bad thing in the case such as this when both models are well formulated. It is usually beneficial to have multiple ways of seeing how things work so as to be able to gain a more complete understanding of how things work.

Chapter Nineteen

The Main Conclusion Is: We Humans Made a Mess on Earth Because the Way We Live Is Incompatible with How Life Works

19.1. The Cause of Incompatibility between How We Live and How Life Works

Basically, the incompatibility is caused by our living mainly our physical part of life while having essentially forgotten we also have our spiritual part of life. Life consists of both parts. Therefore, when we are living mainly our physical part, the way we are living our lives would naturally be incompatible with how things work in the spiritual part of life, which is how things work in the spirit world.

Thus, by being incompatible with how things in the spirit world would be incompatible with how life works. And, by being incompatible with how life works we are making a mess on Earth.

As described earlier in this book, it is like we are trying to run a complex machine with only about half of the knowledge regarding how the machine works. We are bound to make a mess trying to run it.

It is possible that we humans need indisputable evidence that the spiritual part of life exists before we are interested in trying to understand how life works in the spiritual part of life. If this is the case, then consider the following:

1. When we crack open an egg, we cannot tell which part of the ingredients inside would become the wings, which part would become the head, which part would form the legs, etc.

2. However, something somewhere somehow knows how to transform the ingredients into a baby chick.

3. This something is always there and is functioning at the same time the physical part of life is going on. This something is the spiritual part of life. It is what is happening in the spirit world that is going on the same time the physical part of life is going on.

4. The spiritual part of every living thing in the spirit world is directly or indirectly a part of each other. In particular, every human spirit is a part of every other human spirit in the spiritual part of life. And, as explained earlier in this book, the spiritual part of life is more important than the physical part of life. That is because the spiritual part lasts forever whereas the physical part lasts for only short while.

Thus, taking all this into consideration, it makes no sense for us humans to do such things as make wars, mistreat one another, and to do criminal acts of all kinds.

Additional evidence that the spiritual part of life is real and exists is presented in Chapter Two of Reference 1.

19.2. Clear Indications of Incompatibility

If we run the complex machine in a manner incompatible with

how it is supposed to be run, it will run and could make a mess for a while. And, the machine would eventually fail.

Clear indication of how we are living our lives in a manner that is incompatible with how life works is how 87 ancient civilizations covering 3000 BC to 600 AD, a span of 300 years, all eventually failed. This was presented in Reference 9 which describes a study by Luke Kemp, on the 87 civilizations.

Their lifespan of the 87 ancient civilizations ranged from 14 years to 1150 years. So, we humans have been living our lives in a manner incompatible with how life works for thousands of years.

Symptoms of this incompatibility include the following:

1. Early symptoms include wars, mistreatment of one another, and criminal acts of all kinds.

2. Later symptoms include an inability to stop making wars, mistreatment of one another, and criminal acts of all kinds even though we know we could be doing better.

3. Even later symptoms include an increasing disparity between income levels between the well paid and the poorly paid.

4. Very late symptoms include an extreme wide disparity between the few very rich and the rest becoming increasing poor. Homelessness develops, continues to increase, and seems to be uncontrollable. People get increasingly angry, senseless crimes such as mass shootings increase, and a cult mentality seems to develop.

Governing systems become increasingly dysfunctional such that it spends more time being divided and doing infighting than coming together to get things done constructively and meaningfully.

5. Very late symptoms would include how as the population getting increasingly angry it becomes increasingly vulnerable to be manipulated by a poor leader who knows how to manipulate angry people for his own personal advantage. It is well known that angry people could be easily manipulated by bad leaders, Reference 15.

6. An increasing number of people becoming one-dimensional thinkers. Thus, an increasing number of people becoming easily manipulated into doing bad things by a bad leader. A sizable portion of the population would start behaving in a cult-like manner.

Chapter Twenty

A Propose Stepwise Progression for Achieving Spiritual Advancements, and to Help Resolve the Incompatibility

20.1. Achieving Spiritual Advancements Through A Stepwise Progression

Advancements of any kind are usually achieved through stepwise progressions. The knowledge we learned in one step enables us to achieve a stepwise advancement in the next step. Examples of stepwise progressions are our advancements achieved in all forms of transportation, in high technology, in medical technologies, in ecology management, etc.

Stepwise progression is clearly a part of our educational system. Stepwise progression is also a natural part of evolutionary advancements. That was how the human species became the most advanced species of living things on Earth.

In order for the effectiveness of something to be maintained or improved, that something has to be periodically updated and/or supplemented. It has to keep advancing through a stepwise progression in order to stay relevant since everything around it is advancing through stepwise progressions.

Evidence of this is everywhere in our lives, in everything we do, and in everything we use. Anything that does not keep advancing would eventually become irrelevant, ineffective, gradually forgotten, and no longer play a role in our lives.

Up to now our current conventional spiritual models have not been updated or supplemented for thousands of years. Thus, their effectiveness has been slowly declining. As mentioned earlier, church attendance continues to decline, and people are following other models and other criteria in their search for personal spirituality.

The incompleteness and inaccuracies of our current conventional spiritual models was fairly obvious probably from the start, just as the first version of anything would likely be fairly incomplete and not easy to use. Evidence of the incompleteness and inaccuracies in our current spiritual models is their inability to prevent even some Catholic priests from mistreating people, including their mistreatment of indigenous people in the United States and in Canada as described in References 6 and 7.

Spiritual models are among the most important models in our lives since they are a major source of spiritual guidance for our behavior. Yet they are among the few things in our lives that have not advanced and thus their relevance and effectiveness have not been maintained or improved. There appears to be some kind of oversight and a lack of long term long range thinking.

However, it is never too late to start advancing spiritually, and it could be done through a stepwise progression. Stepwise progressions have proven to be highly effective in achieving advancements in every other part of our lives.

Thus, the current conventional spiritual models should not be discarded. Instead they could serve as the first step from which to advance to subsequent steps. The models are basically correct and good, although quite incomplete and inaccurate. But, we have learned from them. Therefore, updating and/or supplementing what we learned should be possible to advance forward onto subsequent steps.

The new spiritual model formulated in References 1 an 2 is more complete and more accurate than our current conventional spiritual models. This is because the way it works is more compatible with how life works. More specifically, its formulation introduced a new way of thinking and new unconventional nonphysical concepts that are more consistent with the nonphysical nature of the spirit world.

Because the spirit world is nonphysical, using nonphysical concepts to model how things work in the spirit world is easier and more effective than using physical concepts to do the modeling as done in developing the current religious models. Physical concepts are inherently inconsistent with the nonphysical nature of the spirit world.

Thus, because the new spiritual model opened up a new ways of thinking and describing how things work in the spirit world, it could reinvigorate our pursuit of spiritual advancement. Up to now, our strictly staying with the conventional spiritual models and their conventional way of thinking has been rather un-invigorating largely because their incompleteness and inaccuracies are becoming increasingly clear.

Our spiritual advancements should have been advancing along with our technical advancements such that their states of advancement are comparable. We should have had spiritual guidance that kept up with our technical advancements as new technical advancements are achieved. .

Up to now, without much spiritual advancements and therefore not much spiritual guidance to help us behave better and wiser we have been applying our technical advancements in bad ways as much as in good ways. Thus, we continue to make a mess on Earth.

This book and References 1 and 2 are written to help stimulate our thinking regarding how we could go about having our spiritual advancement catch up with our technical advancements.

This was also why I continue to encourage others to formulate spiritual models that take science into consideration. The more minds we have working on ways to formulate more complete and more accurate spiritual models, the more ways we will have to view and understand how things work in the spirit world and thus how life works. This would enable us to spiritually advance more broadly, more relevantly, more completely, more accurately, and thus more effectively.

20.2. All Technical Models Combined Would Be Equivalent to a "Model of Everything" for the Physical World. We Should Try Achieving Similarly for the Spirit World

If we think how things work in our physical world is fascinating and amazing, how things work in the spirit world would be way more fascinating and amazing. After all, the spirit world created our physical world and numerous other universes.

Every universe being different and unique is a clear indication the spirit world has to be way more complex and more fascinating than our physical world could ever be.

We talked about pursuing spiritual advancements such that our state of spiritual advancement would be comparable with our state of technical advancements. We have a countless number of detailed physical models describing how a countless number of physical things work in our physical world.

By comparison, we have only a few not very detailed spiritual models describing only how a few things work in the spirit world, and the descriptions are not even very accurate. So, we have long ways to go for our spiritual advancements to catch up our technical advancements.

If we combine all of our detailed physical models, it would be equivalent to being a "model of everything" for our physical world. Ideally, it should be possible to do the same regarding modeling how things work in the spirit world.

But, first we need to regain our awareness of the spiritual part of our lives. This is because to have our spiritual advancements be comparable with our technical advancements we need to be as aware of the spiritual part of life as we are aware of the physical part of life.

I think it is possible for us humans to achieve such awareness because I think we used to have this level of awareness in our far distant past. Indications of this are how various creatures seemed to have retains a lot of their awareness of the spiritual part of life. Rupert Sheldrake, Reference 8, documented a large number of examples of how dogs, cats, horses, etc. appear to have such awareness.

It is possible all living things were created with equal awareness of the spirit world and of the world in which they reside. Therefore, it becomes a matter of retaining or regaining such awareness. The new spiritual model could provide a way to help us do this.

If we could formulate as many detailed spiritual model as have our detailed technical models, then the combination of all of our detailed spiritual models would be equivalent to a model of everything for the spirit world just as how all of our detailed technical models combined would be equivalent to a model of everything for our physical world.

We would then be able to do a lot of things by going through the spirit world much like how we could do a lot of things in our physical world. This might be how some people in our distant past were able to build massive ancient structures such as the pyramids of Egypt. Those people might have been very aware of the spiritual part of life and were thus able to do such things by going through the spirit world.

Chapter Twenty One
Why the Spirit World Won't Come Forth to Solve Our Problems Even if It Could

Ever had a very bad problem and wished the creator (the spirit world) would come forth to help solve the problem, only to find the creator would not come forth? For reasons to be explained, the spirit world really must not do that even if it could.

If the spirit world could solve our problems it would already have the pieces of knowledge we are here on Earth to generate and to be added to the spirit world. The spirit world would then not have a need to create Earth and everything on it in the first place

If a person has a problem and wants help from the spirit world to solve it, and the spirit world was able to do so and gives that person the desired help, the spirit world would then also help every other person who has the same or very similar problem. The spirit world would eventually end up helping everybody since everybody will eventually have the same or similar problem in some moment in life.

If the spirit world were to do this for every problem people on Earth have, this could lead to the spirit world being in danger of failing. How this could happen is as follows.

Earth and everything on it constitutes a representative of a certain portion of the spirit world. Every portion of the spirit world needs to be able to do its part in helping the spirit world maintain its viability. This means each portion of the spirit world must learn and know how to solve issues that could develop in that portion.

Issues are simply another designation for problems, and an issue would develop when something develops an imbalance in its state of knowledge.

Therefore, when a problem develops in a portion of the spirit world, the spirit world would create a universe to help solve it. Thus, it is up to that universe to learn how to solve the problem. The universe would actually be helping the spirit world learn how to solve the problem, because the new pieces of knowledge the universe generates would be added to that portion of the spirit world.

If the spirit world helps that universe solve the problem, assuming the spirit world could, the spirit world would rob that universe the chance to develop the ability for solving problems. That portion of the universe would then not become a viable and productive part of the spirit world. This weakens the entire spirit world's ability to maintain its own viability and thus makes the spirit world in danger of failing.

By analogy, that portion of the spirit world is like a child who must learn how to solve problems in order to become a viable productive adult. If a parent keeps helping the child solve problems instead of letting the child figure out how to solve them, the parent would be robbing the child the chance of developing the ability to solve problems. The child then might not become a viable productive adult.

I have personally observed what could happen when a parent keep helping a child solve his problems. I reminded the parent many times what can happen if it continues, but it continued anyway. She wanted him to have good grades in school. But in my opinion the good grades would be hers, not his, and meanwhile he would not be learning skills he will need in his life.

The child is now in his early 50's. Because he is very weak at solving problems he presents himself as a victim of circumstances to get sympathetic people to voluntarily help him with his problems. For example, he got laid off work because he did not handle his job well. But, he bends the truth by telling people the company went under so everyone there got laid off.

He can't handle stress and heavy workload, and thus can't seem to land and hold on to decent paying jobs. He doesn't manage his finances well and would often need financial help. He doesn't seek self improvement. Instead he spends free time doing fun and pleasurable things.

By analogy, if the spirit world were to help a universe solves its problems (again assuming the spirit world could) the portion of the spirit world the universe represents would not become a viable productive part of the spirit world.

Since everything in the spirit world is directly or indirectly a part of everything else, if this portion of the spirit is lacking the skills it needs to be viable and productive, then the rest of the spirit world would also be lacking those skills it needs as well. This decreases the spirit world's ability to maintain its own viability.

Consequently, the spirit world could be in danger of becoming not viable. This is why the spirit world would not come forth to solve our problems even if it could.

References

1. Book, "Connections with the Spirit World", Second Revised Edition, by Richard Gene, Alive Book Publishing, 2019. (This paper back edition includes a "Preview/Index" to explain various terms and phrases appearing in the text).

Book, "Connections with the Spirit World", by Richard Gene, Alive Book Publishing, 2016 (This hard cover original edition, contains typos that are corrected in the paper back Second Revised Edition).

2. Book, "Understanding the Spirit World & Beyond", by Richard Gene, Alive Book Publishing, 2020.

3. Book, "The Science of Spirituality", by Lee Bladon, Lulu.com, August 1, 2007, and Made in the USA, Coppell, TX on May 2022.

4. Book, "Science and Spirituality, Bridging the Gap", by Kayume Baksh, Austin Macauley, November 30, 2021.

5. Book, "Spiritual Science", by Eric Dubay, Lulu.com, July 12, 2012, and Made in the USA, Las Vegas, NV in May 2022.

6. Article, "Canada: 751 Unmarked Graves Found at Residential School", BBC, June 24, 2021,

7. Article, by the Archdiocese of Toronto "Background for Catholics: Residential Schools", July 9 2021.

8. Book, "Dogs that Know When Their Owners Are Coming Home", by Rupert Sheldrake, Three Rivers Press, 1999.

9. Article, "The Lifespans of Ancient Civilizations", by Luke Kemp of the University of Cambridge, February 20, 2019,

10. "Highly Religious People Struggle the Most with Faith when They Experience Suffering, Study Finds", by Eric W. Dolan, March 22, 2023, PsyPoat, in Mental Health, Psychology of Religion, Social Psychology.

11. Article, "Dark Energy, Dark Matter" author not specified, article appeared on the Internet on March 10, 2023.

12. Article, "A New Study Appears to Stunningly Contradict Newton and Einstein's Theories of Gravity", by Darren Orf, Popular Mechanics, August 14, 2023.

13. Article, "Gravitational Waves Discovered from Colliding Black Holes", by Clara Mackowitz, February 11, 2016, Scientific America Magazine.

14. Article, "Is Artificial Intelligence Advancing Too Quickly? What AI leaders At Google Say", by Scott Pelley, CBS News, "60 Minutes" Sunday Program on April 16, 2023, June 11, 2023, and July 9, 2023.

15. Article, "The Mueller Investigation Revealed that Angry People Are Easy to Manipulate", by John Faithful Hamer, March 27, 2019, American News.

16. Article, "New Research: Diversity + Inclusion = Better Decision Making At Work", by Erik Larson, September 21, 2017, Forbes, Leadership Strategy.

17. Article, "Better Decisions Through Diversity", by Katherine W. Phillips, Katie A. Liljenquist, and Margaret A. Neale, Kellogg Insight, 2009.

18. Article, "How A Diverse Team Drives Better Decision-Making", by Hassina Obaidy, Emtrain, August 16 2019.

19. Article, "Diverse Groups Make Better Decisions", by Heather M. Hill, Physics Today, December 23, 2020.

Index

A

Abstract, abstract thinking, 143, 144
Achievement, 34, 162
Adult, 268
Advance, advanced, advancement, advancements, 38, 44, 68, 90, 101-104, 113, 123, 131, 143, 144, 149, 150, 153, 157-163, 195, 198, 215, 216, 226, 233, 236, 246, 253-255, 261-265
Advance, spiritual advancement, spiritual advancements, 44, 123, 143, 144, 149, 150, 153, 157, 162, 198, 246, 261, 263, 264, 265
Advance, technical advancement, technical advancements, 44, 68, 123, 143, 144, 149, 150, 162, 198, 236, 246, 263-265
Allergies, 37, 161, 162
Analogy, analogous, 29, 57, 82-84, 109, 152, 161, 175, 182, 188, 268, 269
Ancient, 129, 211, 245, 259, 266, 272
Anger, angry, 99, 244, 245, 259, 260, 272
Antimatter, 173, 175
Artificial intelligence, AI, AI's, 225, 226, 228, 230-235, 242
Artist, artists, 221-223
Assemble, assembled, 250

Atom, atoms, atomic, atomic element, 140, 176, 178, 183-187, 256
Attitude, 60
Aura, auras, 111-115, 117, 125, 126, 203, 204, 211

B

Baby, baby chick, 66, 258
Background microwave radiation, 173, 182, 183, 186
Bacteria, 170
Balance, balanced, 54-56, 58, 75, 99, 100, 127, 129, 131, 135-138, 143, 150-153, 157, 161, 187, 191, 223, 239, 241
Baksh, Kayume, 26, 30, 32
Battle, battles, 53, 251
Beginning, 93, 94, 106, 165, 245
Behave, behaves, behavior, 25, 34, 39, 40, 42, 45-47, 58, 59, 73, 90-92, 103, 156, 180, 189, 192, 198, 218-220, 236, 245, 263
Big Bang theory, 173, 175, 177, 182
Bird, birds, 109, 110, 128, 197, 202, 204, 205, 251
Black hole, black holes, 174, 178, 179, 272
Bladon, Lee, 25, 27, 29, 30, 35, 40
Body intelligence, 110
Body parts, 90

Boundary, boundary of a universe, 186, 188-190
Brain, 27, 64, 65, 107-111, 130, 142, 219
Brain intelligence, 110
Building materials, 187

C

Care, care giving, care giver, caring, 59, 64, 69, 70, 72, 73, 112, 137, 146
Cat, cats, 66, 119, 126, 197, 208, 265
Cattle, 66, 197
Chess, 227, 235, 236
Chick, baby chick, 62, 66, 258
Choice, choices, choose, chosen, 41, 70, 74, 92, 116, 117, 153, 155, 234, 242
Church, declining attendance, 45, 262
Civilization, civilizations, 129, 245, 259, 272
Coexist, coexisting universes, 103, 140, 157, 158, 160, 163, 189-191, 215
Colony, colonies, 106, 170
Common, commonly, commonality, 33, 35, 37, 39, 42, 55, 64, 69, 98, 99, 125, 130, 154, 184, 205, 211, 217, 249, 250
Compassion, 137
Compatible, regarding spiritual models relative to spirit world, see also "incompatibility", 29, 33, 36, 76, 91, 144, 174, 263
Compatibilities and incompatibilities, regarding human behavior and spirit world, 57, 63, 99, 100, 129, 135, 137, 141, 150-152, 156, 161, 191, 219, 245, 246, 257-259, 261
Compensate, 108, 154
Complex, complexity, 29, 34, 47, 71, 101, 152, 161, 175, 211, 230, 233, 257, 258, 264
Compose, composer, composers, composition, 131, 163, 221, 223, 226, 227, 236
Computer, computers, computer control center, 90, 107-109
Concept, nonphysical concept, nonphysical concepts, see "Nonphysical concepts"
Concept, physical concept, physical concepts, see "Physical concepts"
Congress, 218
Connection, connections, 78, 82, 83, 88, 95, 96, 100, 111-113, 126-128, 140, 141, 166, 168, 198, 199, 202, 204, 222
Connection, connections of the first kind, 78, 83, 88, 95, 96, 126, 141, 166, 168, 222
Connection, connections of the second kind, 104, 111, 112, 113, 141
Conscious, consciousness, 25-28, 33, 36-38, 40, 62, 64, 73, 80, 85, 86, 90, 94, 107, 109, 111, 115-118, 139, 140, 187, 193, 194, 202, 219, 229-231, 248, 255, 256
Conscious, elemental conscious, 90, 231, 256
Constructive, constructively, 42, 44, 46, 47, 56- 61, 64, 73, 74, 99, 100, 123, 133, 137, 155, 156, 163, 164, 192, 198, 236, 241, 246, 248, 249, 260
Contribute, contributes, contribution, contributions, contributing, 71, 127, 143

Convention, conventional, unconventional, 23, 28, 33-36, 38, 49, 51, 52, 62, 68, 75-77, 81, 80, 86, 87, 91, 92, 141, 144, 175, 183, 184, 220, 262, 263

Correct, corrected, incorrect, correct but incomplete, 43, 45, 52, 75, 90, 144, 162, 220, 262

Create, universes were created for a reason, 53, 54, 131, 135, 191

Creator, creators, 26, 28, 32, 34, 43, 44, 50, 56, 67-72, 75, 77, 88, 89, 94, 95, 100, 122, 139-141, 145, 146, 170, 171, 221, 254, 268

Creation, sudden creation, 105, 215

Creature, 30, 64, 66, 67, 86, 119, 121, 126, 128, 197-199, 202, 207, 211, 218, 265

Crop circle, crop circles, 30, 160, 191

Cross learning, cross-learning, 126, 207, 251

Cycle, cycles, 130

D

Dark matter, dark energy, 139, 140, 173, 180-183, 186

Denomination. Denominations, 92

Design, designs, designed designer, 26, 55, 85, 102, 111, 136, 153, 162, 191, 202, 225-235, 241, 242, 268

Destructive, destruction, 44, 58, 61, 68, 138, 154, 192

Dimension, dimensional dimensions, dimensionality, 29, 55, 76, 97, 112, 185, 189, 190, 231, 233, 236, 240-243, 245, 246, 260

Disease, diseased, diseases, 112, 113, 212, 219

Disparity, 259

Diversity, 60, 61, 81, 122, 123, 127, 155, 247, 248, 250, 251

DNA, DNAs, 26-28, 33, 36, 37, 193-195

Dog, dogs, 66, 197, 206, 265

Doing things by going through the spirit world, 67, 94, 163, 190, 197, 201-203, 207, 210-213, 266

Dolphin, dolphins, 199

Domesticate, domesticated, domestication, domestications, 100, 101, 199

Dominate, dominating, 244

Donate, donated, 124

Donor, donors, 123, 124

Dream, dreams, dreamed, dreaming, dream images, surrealistic dreams, dreaming during restoration, 30, 37, 125, 131, 132, 159, 160

Drown, drown out, 130

Dubay, Eric, 30

Duplicate, duplicates duplicating, duplication, 78, 83, 201

E

Eco, eco system, eco systems, 171

Educate, educated, education, educational, 38, 51, 91, 92, 125, 261

Egg, eggs, 62, 66, 202, 258

Ego, egotistic, egotistical, 65, 70, 72, 98, 132, 133

Egyptian pyramids, 209, 211, 266

Einstein, Albert, 27, 37, 173, 177

Electrical, electricity, 108, 110, 126, 219

Electron, electrons, 23, 139, 140, 176, 178, 186, 187

Elemental, elemental atomic chart, elemental building material, 24, 90, 187, 230, 231, 256

Elemental, see also "Elemental consciousness", "Elemental intelligence", and "Elemental life"

Emerge, emerged, emergence, emerging, 105, 131, 158, 159, 215, 255

Emotion, emotions, 30, 31, 37, 64, 66, 84, 99, 226, 236

Empathy, empathetic, 59, 137

Engineer, engineered, engineering, engineering logic, 26, 27, 34, 35, 38, 76, 81, 82, 86, 162, 171, 201

Entanglement, quantum superposition and entanglement, 24, 28, 37

Environment, environments, 104

Event, events, future events, strange event, strange events, traumatic event, traumatic events, 97, 116, 117, 157, 186, 208, 228, 234

Everyday, everyday experience, everyday life, 33, 39, 49, 61, 54, 83, 125, 171, 217

Evidence, evidences, evident, 26, 43, 64, 117, 155, 158, 205, 251, 257, 258, 261, 262

Evolution, evolutionary, 30, 101, 102, 104, 105, 159, 215, 216, 261

Evolve, evolved, evolves, 67, 101, 102, 131, 158, 161, 203, 216

Express, expressed, expression, expressions, 219, 221, 226, 227, 236

External, externally imposed, 220, 233

Extrapolate, extrapolation, 217

Extrasensory, extrasensory ability, extrasensory abilities, 66, 67, 126, 197-199, 201, 202, 203

F

Fail, failed, failing, 94, 129, 223, 245, 259, 268

Feather, 202

Feature, features, 47, 49, 69, 71, 72, 75, 81, 84-86, 121, 127, 137, 145, 149, 154, 205, 225, 230

Feel, feelings, 51, 64, 84, 99, 117, 125, 130

Fish, fishes, 199

Fission, 183-185

4Qs, 80, 85, 86, 93, 94, 104, 141, 193, 194

Forgive, forgiven, forgiveness, forgiving, 69, 70, 72-74

Foreign, 202, 206

Fossil, fossils, 105, 159, 215, 216

Free, freedom, free will, 54, 69, 136, 150, 153-156, 228, 231, 269

Friend, friendly, friends, friendship, 60, 100, 101, 112, 119, 160, 206, 218

Frontal cortex, 130

Fusion, 183-185

Fulfill, fulfilling our primary purpose, 32, 34, 46, 60, 63, 73, 135, 146, 149-153, 155, 158, 160-163, 191, 223, 248

Future, futures, 37, 74, 96, 105, 116, 128, 129, 140, 204, 228, 233, 234, 236, 237, 242

G

Gap, 23, 26-28, 30, 35, 39, 41, 193

Gas, gaseous, 56, 177, 182

Gene, genes, genetic, genetic engineering, 63, 99, 202, 247
Ghost, ghosts, 30, 65, 119, 120, 126, 208, 209
Giant sea turtle, giant sea turtles, 67
God, perceptions of God, personal God, 43, 51, 52, 72
Google, Google Research Centers, 225, 227
Government, governing system, 219, 243-246, 259
Gradation, gradations, 120-122, 247, 248
Gravity, gravitational wave, quantum particles that might be making up gravity, 55, 76, 173, 174, 177, 179, 182, 183, 186, 188
Group, 207, 250, 251
Grove, grove of trees, 66, 67

H

Hang-up, hang-ups, 244
Heaven, heaven and hell, 40, 69, 70, 74, 144, 145
Helium, 184, 185
Hell, hell and heaven, 40, 70, 71, 74, 144, 145
High tech, high tech device, 38, 207, 261
Highly intelligent living things, 61, 103, 109, 149, 153-155, 158, 160, 161, 163, 188, 189, 191, 192, 199, 211
Hole, black hole, see "Black hole"
Home, homeless, homelessness, 30, 66, 206, 207, 208, 245, 259
Horse, horses, 66, 197, 265
Hospitable, 154, 155, 171
Hostile, 160, 192
Hydrogen, 184, 185

I

Identical, identical twin, 44, 83, 206
Imagine, imagination, imagining, 66, 93, 96, 97, 124, 139, 217, 221
Imbalance, imbalanced, 44, 53-57, 75, 135-138, 152, 157, 223, 268
Imperfection, 43, 44, 89, 93, 241, 242
Impress, impressive, impression, 140
Improve, improved, improvement, 29, 31, 32, 34, 77, 92, 220, 261, 262, 269
In between, 247
In sync, 42, 125
Income disparity, see "Disparity"
Incompatible, regarding spiritual models relative to spirit world, see also "compatibility", 29, 76
Incomplete, incomplete but correct, incompleteness, see "Complete"
Independence, independent, independently, 54, 119, 250
Indigenous people, 43, 262
Ingredient, ingredients, 258
Initiate, initiated, initiation, initial, 80, 81, 86, 87, 93-95, 141, 165, 169, 170, 194, 208, 253
Injure, injury, 108, 219
Inspire, inspired, inspiring, inspiration, self inspiring, 34, 39, 42, 45-47, 64, 68, 90, 192, 219, 220
Instinct, instincts, instinctive, 30, 37, 67, 105, 106, 125, 126, 205

Instruct, instructed, instructs, instruction, instructions, 108, 225-229, 231, 232, 234-237, 241

Intelligence, body intelligence vs. brain intelligence, 110

Intelligence, elemental intelligence, 90, 231, 256

Intent, intention, intentional, intentional, intentionally, 69, 155, 225-232

Interpolate, interpolating, interpolation, 217

Interact, interacted, interacting, interaction, interactions, 38, 80, 85-87, 90, 102, 104, 107, 118, 124, 145, 149-152, 156, 162, 191, 208, 248, 249, 255

Intuition, intuitions, intuitive, intuitively, 30, 37, 81, 96, 106, 107, 125, 126, 132, 205

Invent, invention Inventor, 44, 145, 221, 223, 227

J

Jesus, 210, 211

Judge, judging, judgmental, nonjudgmental, 41, 46, 51, 70, 72-74, 91, 98, 125, 133

K

Kemp, Luke, 129, 259

Kill, killing, 98

Know, all knowing, 32, 43, 69, 70, 72, 90, 154

Knowledge, is analogy with words, 82-84

Knowledge, comes in pieces, 77, 81, 141

Knowledge, is the most powerful thing, 75, 96

L

Language, 69, 71, 72, 119, 226, 232, 236

Laws, laws and rules, 125, 220

Leader, leaders, 219, 243, 246, 260

Life, elemental life, 24, 90, 230

Lifespan, lifetime, 29, 31, 102, 104, 227, 236, 259

Logic, scientific and engineering logic, 27, 34, 35, 38, 76, 81, 82, 162, 171, 201

Long term long range, 262

Loving, 69, 70, 72, 73

M

Major senses, 31, 79, 99, 125, 126, 160, 163, 190, 198, 202

Manipulate, manipulated, 243, 245, 260

Monarch butterfly, 66

Math, mathematical, mathematically, mathematics, 38, 169, 181, 182, 185

Memory, memories, 37, 64, 65, 102, 116, 117

Migrate, migration, 66

Miracle, miracles, 210, 211, 212

Missing link, missing links, 159, 215, 216

Model of everything, 151, 264, 265

Models are essentially documents of our advancements, 143

Models, scientific models that take spirituality into consideration, 25, 34, 35

Models, spiritual models that take science into consideration, 35, 264

Morphic field, 29, 66, 198, 199
Moses, 210, 211
Mother Nature, 104
Motivate, self motivated, self motivating, self motivation, 34, 42, 46, 47
Multi-dimensional, 231
Multi-dimensional way of thinking, 233, 242
Multiple personalities, 116-118
Multiple universes, 29, 31, 158, 162, 190, 216

N

Nation, nations, 163, 206, 219, 243, 246
Natural resources, 162
Near death experience, 30
Nonphysical concepts nonphysical concept, nonphysical concepts, 23, 27, 28, 33, 35, 36, 38, 49, 75-77, 81, 86, 87, 91, 92, 141, 143, 144, 174, 263
Nothingness, 176, 177, 186, 188, 189, 190
Nuclear energy, nuclear reactions, 183-186
Neutron, neutrons, 23, 139, 140, 176, 178, 187

O

Obstacle, obstacles, 39
One-dimensional way of thinking, 231, 233, 236, 240-243, 245, 246, 260
Oneness, 37, 52, 53, 58, 72, 73, 79, 98, 100, 136, 139, 140, 198, 199, 201, 247, 255
Opponent, opponents, 53, 98
Opposite, opposing, 113, 171, 181, 183, 184, 239

Orbit, orbits, orbiting, 186
Organ, organs, vital organ, vital organs, 26, 90, 110, 112, 124, 125, 251
Origin, origin of life, 27, 28, 38, 80, 85-87, 93, 104, 141, 154, 193, 194, 256
Out of body experience, 30, 114, 120, 126

P

Particle, quantum particle, quantum particles, 23-25, 28, 37, 40, 96, 97, 111, 113, 114, 176-184, 186, 188
Partnership, Partnerships, 58-60, 145, 150, 155-157, 222, 223
Past universe, past universes, 37, 131, 159-162, 176, 215
Perception, perception of God, 72
Personality, personalities, multiple personalities, 116-118
Pet, pets, 208, 234
Pilot, piloting, pilots, 58, 108, 109, 110, 142, 150
Physical concept, physical concepts, 25, 26, 27, 29, 30, 31, 33, 36, 40, 41, 76, 91, 140, 144, 145, 263
Physical part of life, 30, 60-63, 65, 67, 68, 73, 85, 103, 121, 133, 137, 141, 144, 145, 149, 150, 156, 201-203, 218, 229, 239, 259, 253-255, 258, 265
Poem, 226, 236
Population, 44, 45, 51, 91, 125, 219, 243-246, 260
Power, powerful, all powerful, 32, 43, 69, 70, 72, 90, 154
Projections, 128, 129, 204, 205, 233
Proton, protons, 23, 139, 140, 176, 178, 187

Purpose, primary purpose, 73, 95, 98, 99, 100, 135, 137, 141, 149, 150-155, 158, 160-163, 191, 223

Purpose, secondary purpose, 73, 96, 98, 99, 100, 135, 137, 141, 149-155, 158, 160-163, 191, 223

Q

Quantum mechanics, quantum physics, 24, 180

Quantum particles, see "Particles"

Quantum superposition and entanglement, 24, 28, 37

R

Recipient, donor recipient, 123, 124

Record of entire lifetime up to the present, 117

Regeneration of body parts, 154

Reincarnate, reincarnation, 29-31, 101, 102, 104, 254, 255

Religion, religious models, 26, 27, 31, 34, 36, 40-45, 52, 69, 71-75, 77, 90-92, 140, 141, 145-147, 162, 254, 263

Remote viewing, 30, 126, 204, 211

Representative of the spirit world, 59, 152, 155, 156, 267

Responsibility, responsibilities, 32, 33, 46, 59, 145, 146, 157, 219, 223, 250

Restoration, sleep is a restoration process, 37, 129-132, 159, 161

Reestablishing reasonable balance, 54-56, 58, 131,135, 153, 223

Retracing, retracing can look like sudden creation, 105, 159, 215, 216

Role model, 70, 71

Rules, see "Laws"

S

Safeguard, safeguards for artificial intelligent things, 228, 232, 233

Safety margin, 154, 155

Salmon, salmons, 67

Self driving vehicles, 242

Senses, see "Major senses" and "Spiritual senses"

Sequence, multiple sequences of universes, 50, 131, 195, 215, 216

Sheldrake, Rupert, 66, 128, 197, 198, 203, 207, 209, 211, 234, 265

Short term short range, 245

Signal, spiritual signals, 67, 79, 84, 108, 110-115, 118, 125, 129-131, 159, 203, 219

Size, size of the spirit vs. size of the brain, 109

Sleep, sleep cycles, 130

Soccer, 226, 227, 235, 236

Something, somewhere, somehow knows how to do something, 62, 80, 88, 158, 222, 258

Soul, 113

Space, space is a physical thing, 173, 174, 176-180, 182-186, 188

Space travel, 163, 188-190, 212

Spirit, 84, 96, 101, 102

Spiritual advancement, 44, 123, 143, 144, 149, 150, 153, 157, 162, 198, 246, 261, 263-265

Spiritual entity, spiritual entities, 78, 79, 86, 88, 96

Spiritual expression, 115
Spiritual guidance, 33, 45, 51, 52, 75, 90, 91, 105, 106, 125, 143, 145-147, 150, 162, 220, 249, 262, 263
Spiritual part of life, 30, 37, 60-68, 73, 74, 85, 98-100, 101, 103, 115, 129, 141, 144, 145, 149, 151, 153, 156, 157, 163, 197-199, 201, 202, 203, 209-211, 213, 217-219, 229, 233, 234, 236, 239, 246, 248, 249, 251, 254, 257, 258, 265, 266
Spiritual senses, 79, 96, 125, 126, 204-206, 221
Stepwise progressions for achieving spiritual advancements, 41, 75, 81, 92, 261, 262
Story, the world of the story, 97, 175
Subatomic particles, see "Particles"
Sudden creation, see also "Retracing"
Surrealistic, surrealistic dream produced during restoration, 37, 125, 131, 132, 159, 160

T

Technical, technical advancement, 38, 44, 68, 123, 143, 144, 149, 150, 162, 198, 236, 246, 263-265
Telepathy, telepathic communication, 30, 66, 125-128, 197, 201, 206-208, 211
Think, thinking, thinking things to exist in our physical world, 96, 97
Thought, thought-like, the structure of every spiritual entity is thought-like, 111,
Three-dimensions, three-dimensional, 97, 112, 185
Tinkering by Mother Nature, 104

Tradition, traditions, traditional, 40
Transplant, transplants, 123-125
Trap, traps, 162, 217
Traumatic events, 116, 117
Travel, see "Space travel" and "Universe travel"
Trees, see "Groves"
Twin, identical twins, 206

U

UAP, unidentified aerial phenomena, 30, 103, 155, 160, 163, 164, 191, 192, 212
UFO, unidentified flying object, 30, 103, 155, 160, 163, 164, 191, 192, 212
United States, 43, 219, 262
Universe, no two universes are the same, 55, 75, 76, 97
Universe travel, 103, 160, 163, 190, 212

V

Virus, viruses, 170
Vulnerable, vulnerability, 94, 260

W

Whale, whales, 110, 199
Win, win-win, win-lose, 98
Word, analogies with pieces of knowledge, see "Analogies" and "Knowledge"

X

Y

Z

Richard holding a photo of late wife Mae and himself at age 40

About the Author

All my life I have wondered what enables life to exist and to function, is there a reason for us to be here, and why am I me and not someone else? When I reached my teen years, I wanted to be able to say on my last day I did my part to make life better for mankind. When I got married and have two kids, I said to myself I really have three kids, the third is mankind, and I will care for my third kid as well as how I care for my two actual kids.

I am unconventional but try to be as conventional as I can so as to be in sync with what it takes to achieve a successful life. However, my unconventional ways still shows up when I am doing something creative. I love to design things and put things together such that things will work smoothly and well and be personalized.

Thus, my wife and I bought the only unsold lot of six lots in a culdesac. It was the most difficult lot to build on, being located on a side of a hill, but it had a magnificent view. I planned the grading and designed the house plan myself. It was so tricky to fit a livable house on the lot that I had to put the garage in the middle of the house instead of at the end of the house, and we ended up with a steep curved driveway.

In my late teens, my mom, one of my sisters, and I designed my parent's new house. Thus, there are two existing houses that I either designed myself or played a part in its design.

After my wife passed away and I retired I turned philanthropic.

I was the first to establish personal endowments at the California Academy of Sciences (CAS) and the Monterey Bay Aquarium (MBA). This led to both places establishing a new program for supports to establish personal endowments. My endowments are to fund educational programs for children and young people, programs that I firmly believe would help children and young people become better adults. I also help funded to a similar educational program at the Ruth Bancroft Gardens (RBG). The CAS awarded me their "Spirit of Philanthropy Award" for year 2022. I am also in the process of establishing an endowment at UCSF to support their internship programs for high school students.

All these mentioned educational programs for children and young people increase their knowledge regarding how life works. My work in writing my books does the same. Thus, a strong connection exists between my work and the work of these institutions. Therefore, by writing my books and supporting these institutions I am carrying out my work on two different paths.

The three books I wrote was formulate and present a new spiritual model that models how life works more completely and more accurately than the spiritual models that exist up to now. The purpose is help mankind gain more completely and more accurately the knowledge of how life works. Therefore, mankind would be more able to form the wisdom to stop making wars, mistreating one another, and doing criminal acts of all kinds. The new spiritual model shows how it is that our creator did not mean for mankind to behave so poorly. Instead, the model revealed mankind has a primary purpose for being here on Earth, and the way to fulfill this primary purpose is to behave in a manner that is constructive and meaningful.

Self portrait oil painting by Richard at age 30

ABOOKS

ALIVE Book Publishing and ALIVE Publishing Group
are imprints of Advanced Publishing LLC,
3200 A Danville Blvd., Suite 204, Alamo, California 94507

Telephone: 925.837.7303
alivebookpublishing.com